FOREST PRODUCTS

Advanced Technologies
and Economic Analyses

Academic Press Rapid Manuscript Reproduction

FOREST PRODUCTS

Advanced Technologies and Economic Analyses

David A. Tillman

Principal Scientist
Envirosphere Company
Ebasco Services, Inc.
Bellevue, Washington

1985

ACADEMIC PRESS, INC.
(Harcourt Brace Jovanovich, Publishers)
Orlando   San Diego   New York   London
Toronto   Montreal   Sydney   Tokyo

ACADEMIC PRESS, INC.
Orlando, Florida 32887

United Kingdom Edition published by
ACADEMIC PRESS INC. (LONDON) LTD.
24–28 Oval Road, London NW1 7DX

LIBRARY OF CONGRESS CATALOG CARD NUMBER: 85-70363

ISBN 0-12-691270-X

PRINTED IN THE UNITED STATES OF AMERICA

85 86 87 88   9 8 7 6 5 4 3 2 1

For Millie

Contents

Chapter III. The Discount Rate for Current Forest Industry Investments

Chapter IV. Discount Rates for New Technologies in the Forest Products Industry

Chapter V.   Advanced Pulping Processes

Chapter VI.   Advanced Processes for Lumber Manufacturing

Preface

In 1981–1982 Amadeo Rossi, Stephen Simmons, and I performed an analysis of wood processing technologies for the U.S. Congress, Office of Technology Assessment. We reviewed over 100 present or potential innovations applicable to the Forest Products Industry. Since that time the industry has emerged from its worst depression since the 1930s. At the same time, it continues to show strains associated with certain fundamental problems. Now the lumber manufacturing segment of the industry faces at best a stagnant market, overcapacity, and substantial intermaterial competition. Plywood, as a broad based commodity, is a product whose time has come and gone. Pulp enjoys a robust market now, but is caught in a severe cost/price squeeze.

The Forest Products Industry now faces additional supreme challenges in converting from an old growth to a second growth resource base, and from a diverse to a concentrated industry. These challenges include the perception by investors that the Forest Products Industry has become increasingly risky. The consequence of that risk perception is that capital is increasingly expensive for wood products manufacturers.

The Forest Products Industry faces these challenges with an arsenal of potential new technologies. These technologies can be used to improve the economics of existing major products such as pulp or dimension lumber. They also have the potential for bringing wood into such new market arenas as energy, chemicals, and foodstuffs. Application of the best of these technologies can aid in converting forest products firms from a stagnant smokestack industry into a more vigorous, growing enterprise. In such a conversion, the perceived risk associated with the industry may be reduced by penetration in diverse and countercyclic markets.

The number of potential new technologies within the Forest Products Industry is considerable. There is, in fact, a cornucopia of concepts for improving methods to produce pulp, lumber, panel products, fuels, electricity, dietetic foods, etc. Such ideas range from minor improvements within existing technologies to major departures and production processes. Some ideas have a sense of urgency and immediacy in terms of installation and adoption. Some ideas have broad applicability.

This monograph is designed to address certain issues associated with such technological advancement in the Forest Products Industry. In addressing those advanced technologies, this book provides the following: (1) a review of the state of the industry, (2) a proposed technique for innovation analysis and the basic parameters necessary for such analysis, (3) a review of many of the possible innovations available to the forest industry, and (4) detailed evaluations of a few selected potential innovations.

The writing of this monograph began by reexamining many of the most promising innovations evaluated for the Office of Technology Assessment. Economic and more detailed technical information were added to the data base, as the OTA analyses were strictly material and energy balances. New technologies that had not emerged at the time of the OTA study were also evaluated. Such new technologies included microfibrillated cellulose, an invention of the Rayonier Division of ITT. Such innovations provided the basis for this volume.

The completion of this monograph was accomplished with considerable assistance from numerous individuals. William Aho of Ekono, Inc. provided much of the data base on autocausticizing and organisolv pulping. Ramsey Smith of the University of Washington and Don Arganbright of the University of California provided much information on lumber products and processes. David Augustine of Ebasco Business Consulting Company provided assistance and literature on the financial aspects of the analysis. Mike Meagher provided computer programming support. Bob Keegan of Ebasco Services and Bob Kundrot of Koppers Company along with Ramsey Smith, Bill Aho, and Dave Augustine provided reviews. Special acknowledgment should go to my wife, Millie, who handled manuscript preparation.

This monograph, then, offers one view of potential technological futures for the Forest Products Industry. It is not a prescription for success, but it is an analytical approach to the problem of revitalizing an industry. It offers a means for considering innovations that provide investors unusual returns for the risk taken. It provides analysis of a few such innovations.

Chapter I

THE CHALLENGE OF CHANGE IN THE FOREST
PRODUCTS INDUSTRY

I. INTRODUCTION

The Forest Products Industry (FPI) faces severe challenges
at this juncture in its history. It must convert its business strategy from a resource orientation to a marketing
orientation. At the same time the FPI must convert its raw
material base from a naturally occurring resource to a managed resource, and it must convert its technical framework
from large log to small log oriented technologies.
The wood products industry must recognize that it is an
integral part of a much larger materials community which includes ferrous and nonferrous metals companies, nonmetallic
mineral corporations, and petrochemical firms. The FPI must
compete aggressively with these suppliers of materials used
by society. No longer can the FPI stand pat on the statement
"wood is good," and remain aloof from the real competitive
forces of the marketplace.
The challenges facing the FPI are formidable but not in-
surmountable. The forest industry has faced numerous traumatic events in its history and has overcome serious economic
obstacles successfully. In the period 1820-1870 the FPI lost
the charcoal market to coal based coke [39]. The FPI lost
dominance of the entire energy market, when coal became king,
in the period 1860-1900 [11]. Coal displaced wood as the
primary feedstock for chemical manufacturing during the
period 1920-1940 [6].
The loss of energy and chemical markets during the 19th
Century and the early years of the 20th Century was counter-
balanced by the invention of mechanical and chemical pulping.
During the period 1850-1900 most of the major processes for
wood pulping were invented: stone groundwood, kraft, soda,
and sulfite [23]. Entire new markets for wood products
emerged. During the 1950s the FPI introduced the concept of
pulping sawmill residues [15]. During the late 1940s and
1950s the FPI began producing dry particleboard from

residues [24]. Resource conservation, and more complete
resource utilization, became reality.

During the 1950s the FPI changed from a regionally
oriented industry to a series of national firms. At this
time the concept of vertical and horizontal integration took
hold, and firms began producing the full range of forest products. During the 1960s the integration of lumber and pulp
or particleboard production on a single site gained momentum.
This concept of mill integration, which has existed for many
decades, has become a powerful force in the production
process.

Certainly technical, marketing, and managerial innovations have occurred in the FPI throughout its history. The
overriding challenge facing the industry today is to acceler-
ate the current pace of innovation. Unless more emphasis is
placed on innovation, the FPI will continue to lose economic
ground to more aggressive competitors.

In order to understand the current challenge of change,
and the potential technoeconomic responses to that challenge,
it is useful to examine the following issues: (1) the current
production and marketing structure of the FPI; (2) the current
economic structure, and the financial (investor oriented) performance of the FPI; and (3) opportunities for improvement
in the economic performance of the industry through technological innovation. While these issues are addressed in detail throughout the text, they are summarized in the sections
below.

II. CURRENT PRODUCTS AND MARKETS OF THE FOREST PRODUCTS INDUSTRY

A market oriented firm approaches the marketplace by asking what the market demands, and how that firm can fulfill
market demand at a competitive advantage. A product oriented
firm approaches the marketplace by defining its product slate
and selling that particular slate in the marketplace. In
general, the FPI is not market oriented. This industry is
product oriented. The FPI focuses its technical efforts upon
its present product line, making incremental improvements in
the costs of producing such goods in order to maintain or
boost sales.

The FPI produces the following basic goods: (1) lumber,
(2) veneer and plywood, (3) panel and composite products,
(4) pulp based products, (5) fuels and electricity, (6) specialty and extractive based chemicals, and (7) miscellaneous
minor products (e.g., beauty bark, poultry bedding). The FPI

also includes numerous secondary manufacturing firms making such goods as furniture and baseball bats.

Of the major product areas, pulp and lumber are dominant, as they command over 80% of the total market for wood based materials (see Table I). Further, they face the most severe competition from mineral based commodities and products. Consequently they are the focus for much of the remaining discussion in this chapter and in this text as a whole. Plywood and composites, including engineered waferboard (or oriented strand board) appear to be locked in an internal struggle, with waferboard beginning to win handily. Energy and chemicals are minor but significant markets for the FPI.

A. Lumber and Board Production and Markets

The sawmill is perhaps the most basic production unit of the FPI. Its outputs are used directly as products and also as raw materials for the manufacture of furniture, specialty products, and (because residuals such as chips have achieved product status), pulp and particleboard. Residential housing is the dominant market served by the lumber and board industry, with secondary markets including housing renovation, commercial construction, railroad ties, and posts and poles.

1. Sawmill Production Technology. In general, the purpose of the sawmill is to fit round pegs (logs) into rectalinear, if not square, shapes (boards, studs). To accomplish this task the sawmill subjects the log to a variety of operations as shown in Fig. 1. Figure 1 demonstrates that the critical operations or machining centers in the sawmill include (1) merchandising and bucking, (2) debarking, (3) the headrig, (4) cant breakdown, (5) edging, and (6) trimming. Drying and planing are unit operations that may be added depending upon market conditions and requirements. Of these machine centers, only the dry kiln does not involve subjecting the piece to some mechanical action in order to alter its size and/or shape.

On average, the modern sawmill now recovers about 40% of the small log (e.g., 8-12 in. diameter, large end) in the form of lumber [19]. Actual mill-specific yields depend upon detailed conditions of the raw material (e.g., sweep), and specific sawmill technology employed. Beyond lumber recovery, some 30 to 35% of the log can end up as pulp chips. In the current mills as little as 15 to 25% of the logs fed to the mill end up as low value residuals (e.g., fuel).

2. The Market for Sawmill Products. The two primary products of the sawmill are lumber and pulp chips. The new

TABLE I. *U. S. Production of Major Wood Products, 1950– 1981*

| | Product | | | | | | |
| Year | Lumber | | Plywood and Veneer | | Panel products[a] | | Pulp products | |
	Tons × 10^6	%	Tons × 10^6	%	Tons × 10^6	%	Tons × 10^6	%
1950	42.6	71.1	2.1	3.5	1.3	2.2	13.9	23.2
1955	41.6	62.7	3.6	5.4	1.7	2.6	19.5	29.4
1960	36.0	53.7	4.7	7.0	2.2	3.3	24.1	36.0
1965	40.6	48.2	7.7	9.1	3.5	4.2	32.5	38.6
1970	38.3	40.8	8.6	9.2	5.3	5.6	41.7	44.4
1975	35.5	38.5	8.9	9.7	7.1	7.7	40.7	44.1
1979	41.3	37.8	10.5	9.6	9.6	8.8	48.0	43.9
1981	31.8	27.9	8.5	7.5	9.1	8.0	64.4	56.6

[a]*Includes composites.*

Sources: Ulrich, 1981, U.S.D.A. Forest Service [37]; U.S. Dept. of Commerce, 1934 [38].

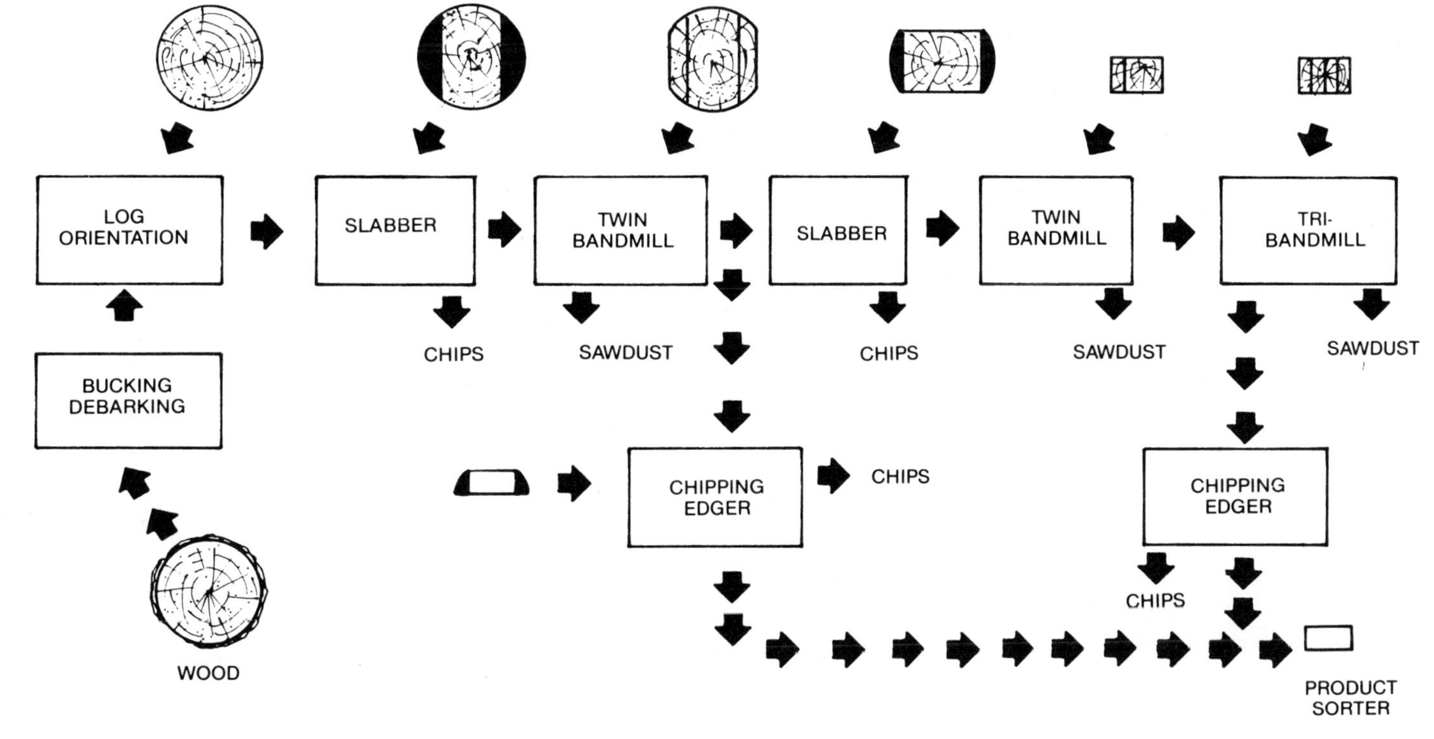

FIGURE 1. *Material flows in a typical small log sawmill. Source: [34].*

residential housing market is the primary consumer of lumber while kraft and thermomechanical pulp mills are the primary consumers of chips. Because the lumber market provides greater revenues for sawmills, and because the pulp market is discussed in a subsequent section, the lumber and board market is discussed in this section.

The sawmill is becoming progressively more efficient in converting logs into lumber; however the housing industry also is becoming more efficient in the use of that lumber. This combined efficiency of production and use has contributed to the current situation where the solid wood products industry can support the construction of 3 million homes per year while the market demands little more than half of that capacity, even in a good year [28].

The housing market is becoming more efficient in its use of lumber principally by two mechanisms: (1) building more and more mobile homes and multiple family dwellings (e.g., condominiums) at the expense of more lumber intensive single family homes; and (2) decreasing the lumber component per square foot of living space for all types of homes.

The first mechanism is shown by Table II, and by Figs. 2 and 3. The second mechanism is illustrated by Fig. 4. Of these mechanisms or trends, the shift in housing type is most dramatic. Figure 3, which depicts the percentage of

TABLE II. Housing Starts by Type
(Values in units × 10^6)

Year	Single family homes	Multi family homes	Mobile homes	Total
		Housing starts		
1953	1.24	0.22	0.08	1.54
1957	1.05	0.20	0.12	1.37
1961	0.99	0.38	0.09	1.45
1965	0.96	0.51	0.22	1.69
1969	0.81	0.66	0.41	1.88
1973	1.13	0.91	0.57	2.61
1977	1.45	0.54	0.28	2.26
1981	0.70	0.38	0.22	1.31

Source: Spelter and Phelps [30].

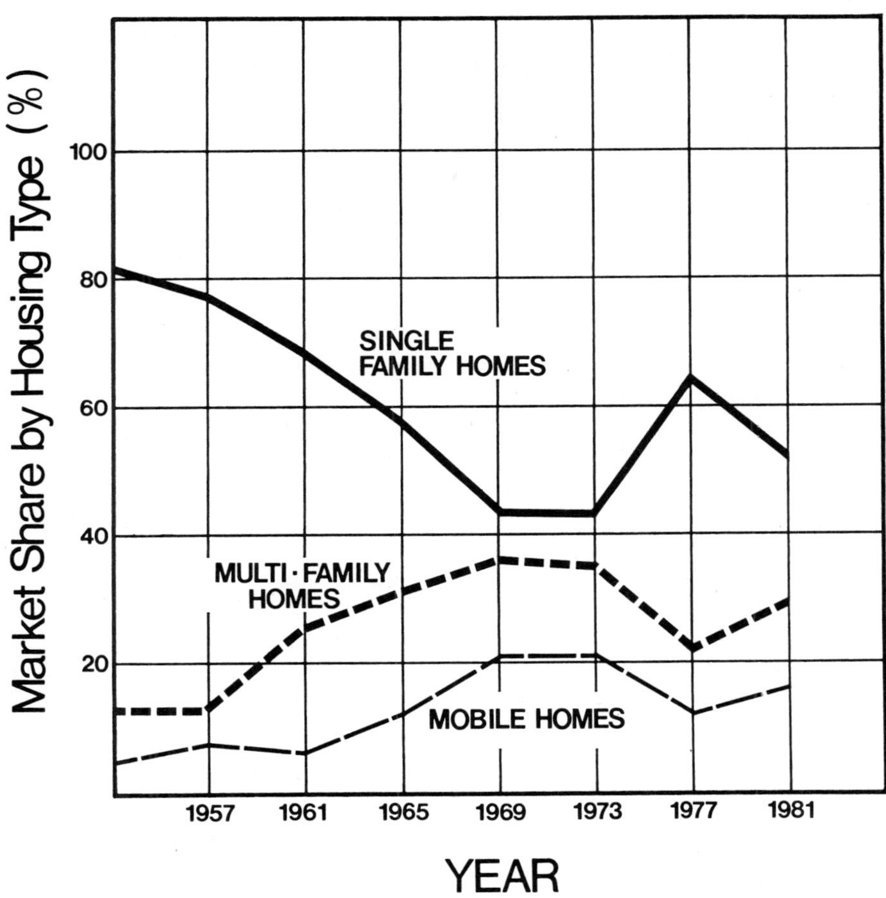

FIGURE 2. Distribution of the housing market by type
of home. Source: [30].

mortgages made by Weyerhaeuser Co. by type of home, is most
dramatic. This trend towards multi-family structures is
likely to continue for the foreseeable future due to the
demographic forces existing in the U.S. Single family homes
do appear to be getting somewhat larger [30]. This increasing size of single family homes is the only factor that mitigates at all against a potential long term lumber market con-
traction.

Not only does the FPI face a lumber market that may be
contracting, but it also faces severe competition from other
segments of the materials community for that market. Framing
lumber and siding must compete against steel, aluminum, and

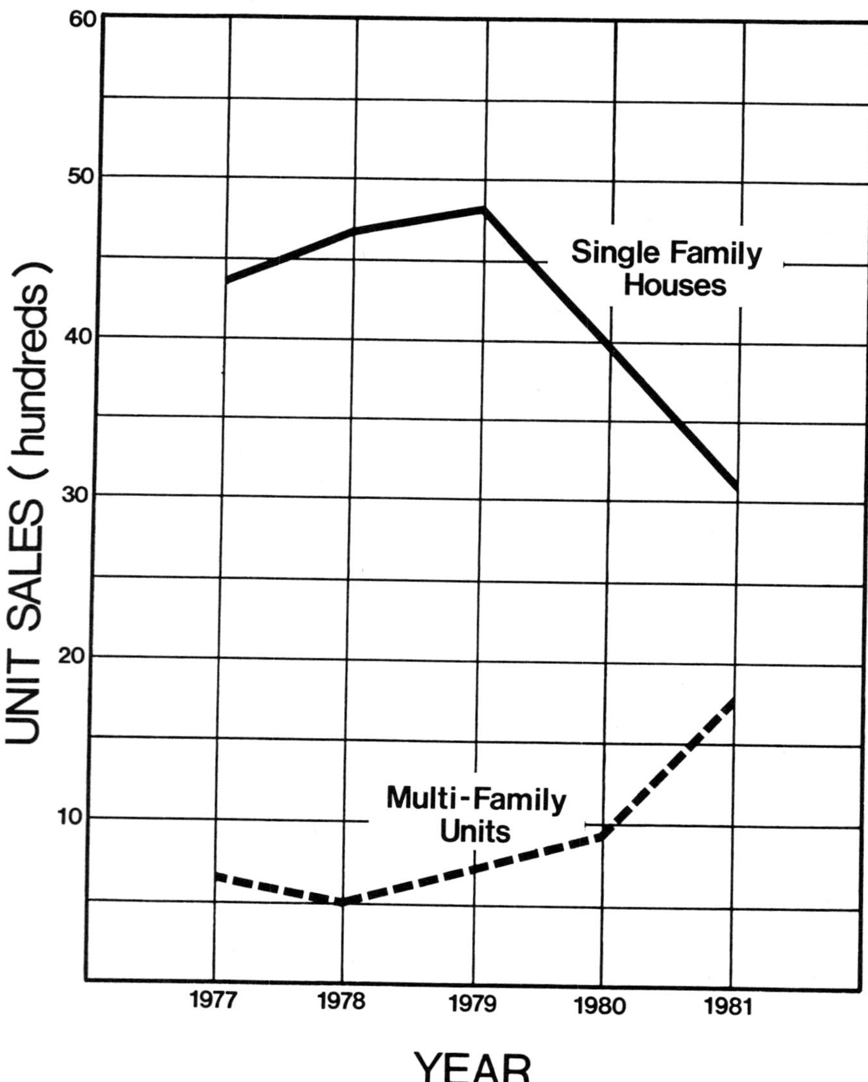

FIGURE 3. Unit sales of homes by type, financed by the Weyerhaeuser Co., reported in the 1981 Weyerhaeuser Co. Annual Report.

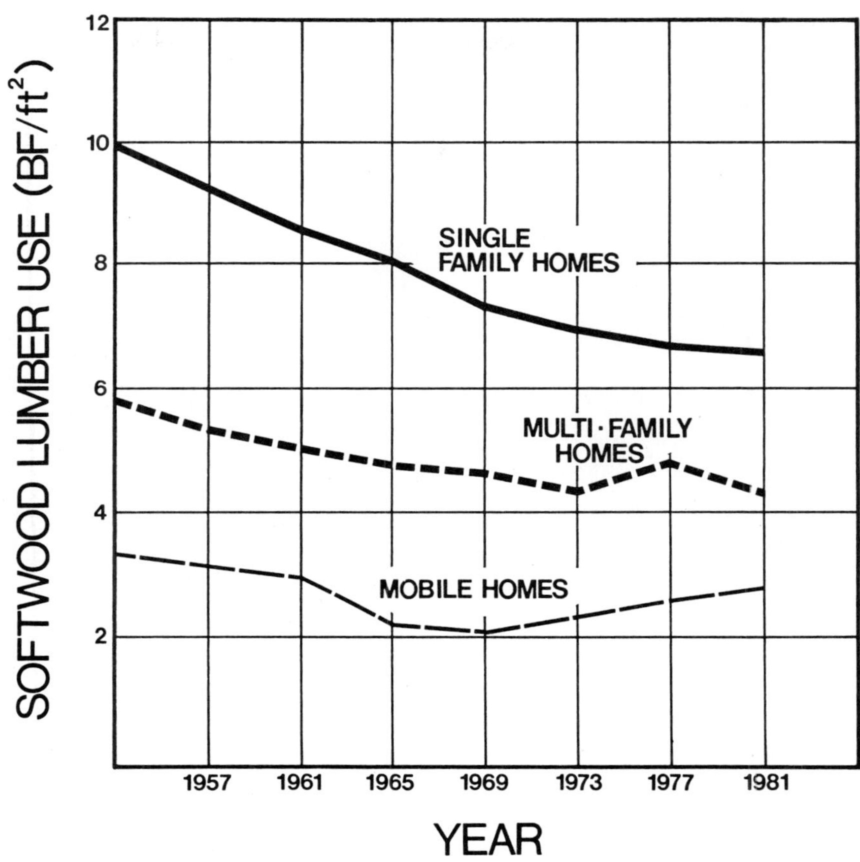

FIGURE 4. Intensity of lumber use by type of housing
structure, measured in bd ft of lumber per ft² of
housing structure. Source: [30].

plastic substitutes. This competition is depicted in Figs. 5
and 6. Figure 5 is a single family home being framed with
aluminum studs advertised to be more termite and rot resistant than wood studs. Metal studs are expanding their market beyond commercial structures (e.g., office buildings),
where they now dominate. Figure 6 is a series of condominiums with vinyl (plastic) siding.

Only those firms with highly efficient mills or market
specialties can remain profitable in this highly competitive
marketplace. The sawmill capacity lost during the 1981-1983
recession may never be replaced, and more capacity may be

FIGURE 5.   A single family house being framed with aluninum studs rather than wood studs.

FIGURE 6.   A condominium development using vinyl siding rather than wood siding.

lost in the ensuing years. Survival in this market demands innovative attention to technologies that achieve additional dramatic cost reductions, even over the current production methods.

   3. Applicability of Lumber Data to Other Solid Wood and Board Products. Because sawmills are such central systems in the FPI, they provide the obvious focus for production and market analysis. The situation is far more bleak for plywood than for lumber. Plywood is a product whose time may have come and gone, at least as a bulk commodity. Plywood may become a specialty product in the near future. Other, less significant, product areas face similar problems including: (1) cut stock and household trim (e.g., window frames, exterior doors, and interior moulding); (2) furniture manufacture (particularly office furniture, tables, and chairs); (3) specialty products (e.g., pencils, toothpicks, broom handles, baseball bats, tennis rackets, and picture frames).
   These secondary products are well publicized areas of competition between wood and mineral based substitutes. Such competition has existed for some time. From youth leagues to the college World Series, coaches have learned to grimace and holler "run at the ping of the bat!" Yogi Berra stories about reading the trademark to avoid breaking the bat have no meaning to younger, amateur baseball players. The aluminum bat has all but eliminated its wood counterpart. Such successes by mineral based products do not bode well for solid wood products.

B. Pulp Production and Markets

   Wood pulps and their consequent products are the most significant products made by the FPI today. More tons of product, and dollars of revenue, are generated as a result of pulp mill activities than the activities of any other manufacturing system of the FPI including lumber production. Further, pulp mills not only generate products and revenues, but they also serve as markets for high value byproducts of the lumber and plywood manufacturing systems.
   To accomplish the revenue generation now being achieved by pulp manufacturing, wood pulps are used in the final manufacture of such products as newsprint, printing and writing papers, gray and bleached boxboards, fluff pulps (e.g., for diapers), dissolving pulps and their products (e.g., rayon, cellophane), corrugated containers, construction papers, bags, and a host of other commodities.

1. Pulp Manufacturing Processes. Two types of pulping processes exist: mechanical pulping and chemical pulping. Mechanical pulping includes stone groundwood (SGW) pulping, refiner mechanical pulping (RMP), thermomechanical pulping (TMP), and recycle paper pulping. Chemical pulping includes the kraft or sulfate process, sulfite pulping, soda pulping, and neutral sulfite semichemical (NSSC) pulping.

a. Mechanical pulping processes. Mechanical pulping processes are the oldest methods for liberating fibers from wood, and they involve reducing furnish to fiber by one or another form of grinding or beating. Typically mechanical pulping processes have yields in the 85 to 95% region.

Mechanical pulping systems can be distinguished one from another by furnish type, specific energy consumption per unit of pulp produced (e.g., kWh/ton), and by pulp product strength. SWG pulping uses logs as furnish while RMP and TMP use chips (including sawmill residuals). SGW pulping uses about 1800 kWh to produce one ton of pulp while RMP systems use about 2200 kWh/ton and TMP systems use 2200-2700 kWh/ton depending upon species and the degree of refining [34].

TMP consumes residuals as furnish, and produces the strongest mechanical pulp with the highest percentage of long fibers. As a result, TMP pulping has achieved the broadest product flexibility, and has become the state of the art in mechanical pulping [3]. Its generic flowsheet is shown in Fig. 7, and its use is shown in Fig. 8.

Recycling, either of newsprint or old corrugated containers, uses previously processed wood pulp as furnish. Recycling is characterized by having the lowest cost furnish, the lowest specific energy consumption, and very high yields. Economically recycling is a highly desirable method of mechanical pulping, being competitive with integrated virgin pulp and paper manufacturing operations [35].

All mechanical pulps exhibit a common property: strength limitation. Because they are produced by attrition of the furnish, the individual fibers retain their lignin content. These fibers are relatively stiff, and the consequent pulp sheets therefore have a limited number of contact points between the individual fibers. This limited contact causes the relatively low strength of mechanical pulps (relative to chemical pulps such as kraft pulp).

b. Chemical pulping processes. Kraft pulping is by far the dominant form of chemical pulping, and all wood pulping. Its popularity results from its early development as a pulping system with the ability to use virtually any species of wood. More significantly, kraft pulping processes accomplish recovery of pulping chemicals in a straightforward manner.

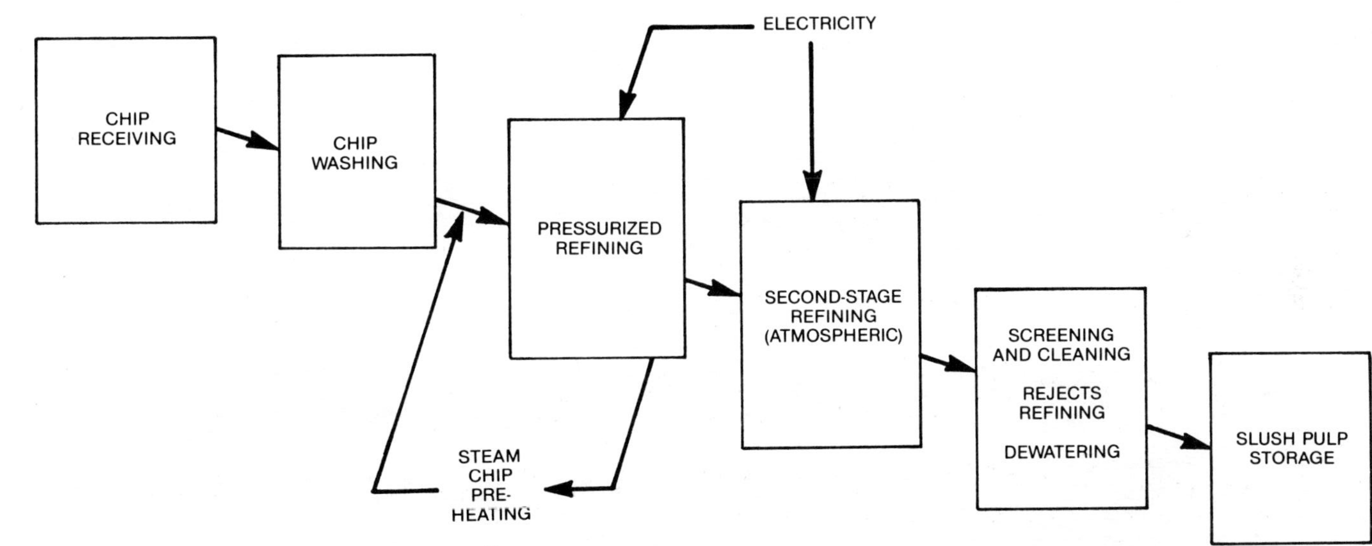

FIGURE 7. Generic flowsheet of a TMP pulp mill.

FIGURE 8. The NORPAC newsprint machine. (Photo courtesy of Weyerhaeuser Co.)

Finally this process produces a product of high strength and consequent utility. Like all chemical pulping processes, kraft pulping removes lignin and some hemicelluloses from the wood, thus flexibilizing the fibers while liberating them from each other. This lignin removal results in the high strength properties of kraft pulp [1, 34].

The kraft process employs NaOH and Na_2S as the active chemicals, and cooks the pulp chips in these chemicals at a pH of 13-14 for a period of 60-90 min. and at 320-360°F (160-180°C)[1]. The generic process for kraft pulping is shown in Fig. 9. Yields in the kraft process depend upon the desired characteristics of the pulp produced, hence the type of wood employed (e.g., hardwood vs. softwood), the length and severity of the "cook," and the degree of bleaching (secondary treatment of the pulp with some combination of chlorine, chlorine dioxide, oxygen, hypochlorite, and/or extraction processes). Typical kraft pulp yields are 45-50%. The highest yields are associated with unbleached kraft pulping of hardwoods. The highest strength pulps come from bleached kraft produced from softwoods [1].

Kraft pulping, like all chemical pulping, requires a combination of thermal and electrical (mechanical) energy. Typical energy consumption levels are $12\text{-}15 \times 10^6$ Btu (thermal) and 850 kWh per ton of pulp [1, 18]. Most, if not all of this energy can be supplied from internal energy sources as a result of bark and spent pulping liquor combustion [8, 18].

2. Markets for Pulp Products. Pulp penetrates a wide variety of markets, where it competes with such products as film plastic. The most visible competition occurs in the area of packaging, where pulp based products compete with petrochemical products such as styrofoam, polyethylene, polystyrene, and other engineered polymers. That visibility is caused by the use of packaging as an advertising medium. Products associated with that competition include milk cartons, egg cartons, meat trays, grocery store bags, department store bags, 6-pack containers for beer, and a host of other items as is shown in Figs. 10-12. That competition exists because of the relatively high costs associated with pulp production [1]. That competition is discussed in detail by Wolpert [40], who highlights the product advantages of both petrochemical and cellulosic products.

While competition among packaging materials presents serious problems, efficiency in the use of packaging may present a more subtle change. Over 90% of all goods in the U.S. are shipped in corrugated containers. A change in the ASTM standard for such containers promises to reduce the corrugated container material requirement by 23% [20]. This

                                                    DAVID A. TILLMAN

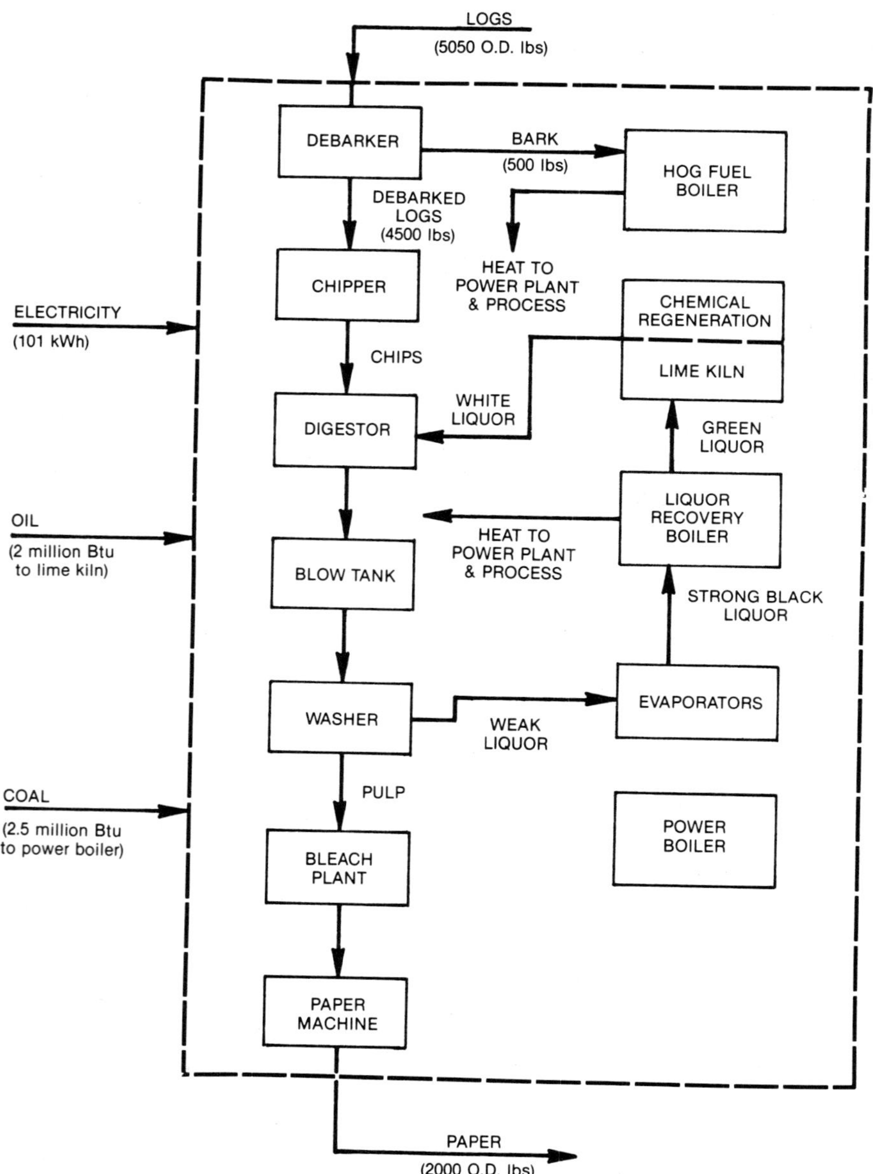

FIGURE 9. *Generic flowsheet of a bleached kraft pulp mill.*

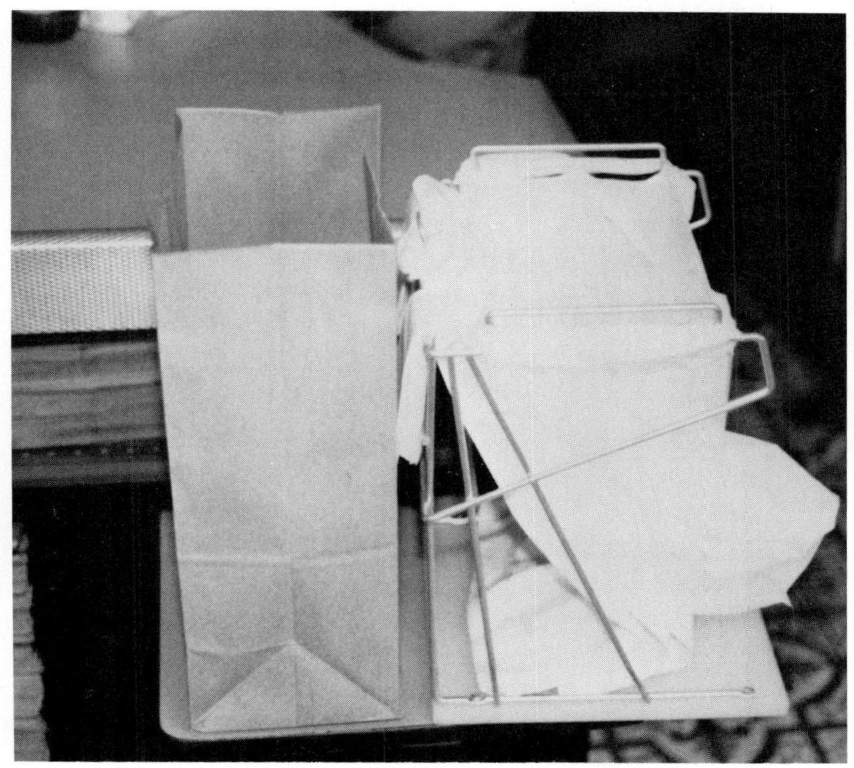

FIGURE 10. Competition between kraft bags and plastic bags in the grocery store. (Photo courtesy of Shop Rite Grocery, Kirkland, Washington.)

change is based upon different testing methods. It has pro-
found implications for the FPI in two directions: (1) retaining the competitive position of the FPI in this market
by reducing packaging costs per unit of product transported;
and (2) reducing the volume of corrugated containers (hence
linerboard and corrugated medium) required, and therefore
decreasing sales and profits associated with this product
area.

Competition and efficiency characterize the packaging
market. Other markets, however, are characterized by product
complements rather than competitors. Such is the case in
printing and writing papers, and newsprint. Projected competition has not materialized. The computer, including the
word processor, was projected as a competitor for paper.
This competition was based upon the projected electronic

FIGURE 11. Competition between kraft milk cartons and plastic milk jugs in the grocery store. (Photo courtesy of Shop Rite Grocery, Kirkland, Washington.)

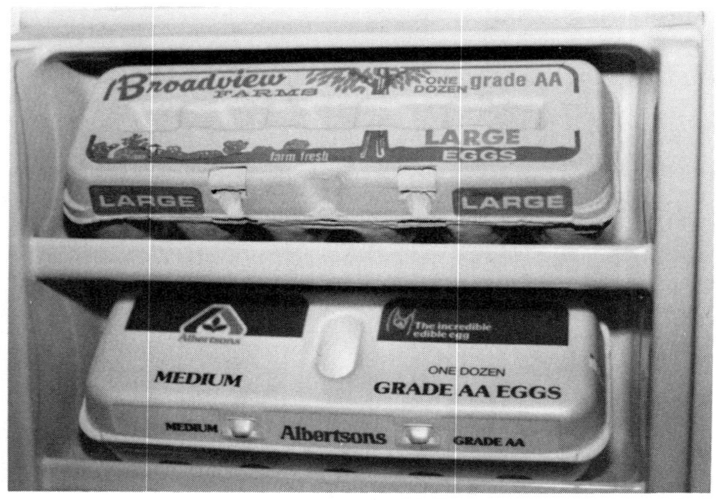

FIGURE 12. Competition between pulp and styrofoam egg cartons.

office. Office networks connected by computers and word
processors were assumed to have less demand for paper. Data
storage and transfer could be accomplished without paper.

This competition has not materialized. The personal
computer (installed either in the office or the home) and the
mini-computer, along with the mainframe machines, have
created an enormous demand for paper. Accounting and financial reports previously considered too laborious and costly
to develop now are created routinely, and converted to "hard
copy" for management perusal. Intermediate drafts of technical reports, professional papers, and books now proliferate.
Catalogues and direct mail advertisements hawking the wares
of the fiercely competitive computer hardware and software
vendors abound.

Electronic systems and paper are not totally without competition. In past decades many metropolitan newspapers have
ceased operations. At the same time network television has
developed extensive news services while the Cable News Network has emerged. Other potentially competitive television
services are being promoted. Simultaneously, however, the
explosion of information has led to the creation of many new
magazines and professional journals. Whether electronic news
and information services and channels will reduce or increase
the long run demand for paper (including newsprint) remains a
question.

In the short run, then, the market for pulp based products
remains strong [3]. For the remainder of this century, however, pulp based packaging products will continue to face
price and convenience competition in the supermarket and the
department store. Further, they will face trends of increasing efficiency in the use of packaging materials. When one
type of material, linerboard, comprises 25% of all pulp based
products [20], such problems cannot be ignored.

The pulp industry does not face the severe problems of
overcapacity associated with the lumber and board industries.
The pulp industry, however, faces a more severely competitive
marketplace where competitors are investing heavily in their
future, in research and development dollars. Unless the pulp
and paper industry adopts innovations at a rate that is competitive with the rates experienced by the petrochemical industry, its long term market position may decline.

C. Other Markets and Products for the FPI

Lumber has been taken as a proxy for all board products,
while pulp has been used implicitly to represent most chemical products. The latter representation may be used because
many wood chemical products (e.g., rayon) are made from

dissolving pulp while others are produced as byproducts of
the various chemical pulping processes. A partial list of
such byproduct chemicals includes tall oil, lignosulfonates,
dimethyl sulfoxide (DMSO), and torula yeast. This byproduct
relationship is elucidated by Frederick [12]. Energy, chemical, and related products afford the FPI with a major growth
opportunity. These ancilliary products include fuel, electricity, beauty bark, extractive chemicals (e.g., turpentine),
lignosulfonates, DMSO, cattle food supplements, drilling mud
additives, and the like. Of these alternatives, energy products have the most economic significance. Of particular
interest is electricity as produced under the provisions of
the Public Utility Regulatory Policies Act (PURPA). PURPA
requires utilities to purchase electricity produced by cogenerators at the marginal cost that would otherwise be incurred
by utilities in generating that power.

Because cogenerators can sell their power at the marginal
cost of electricity generation (usually defined as the full
avoided cost), this practice has increased dramatically.
Cogeneration now supplies 7% of the electricity consumed in
the U.S. (as of 1984). In 1980 cogeneration supplied only
3% of the electricity used in this country [7]. The FPI always has been a leader in cogeneration, supplying more power
by this mechanism than any other industry prior to the advent of PURPA [33]. While the practice of cogeneration was
always strong in pulp and paper prior to 1978 and the advent
of PURPA, this legislation gave substantial impetus to its
use in lumber, board, and panel manufacturing [13]. Further,
more nonintegrated paper mills and paper converters are now
using cogeneration.

The FPI uses the conventional steam topping (Rankine)
cycle for cogeneration. The steam topping cycle is most
applicable for solid fuel systems. Further, the Rankine
cycle is the most thermodynamically efficient approach to the
simultaneous production of electricity and process heat [32].
Its limitation is a low electricity/process heat ratio relative to other cogeneration cycles (e.g., Brayton or Diesel
cycles).

Numerous improvements have been made in the use of the
Rankine cogeneration cycle, including the use of higher steam
pressures at the inlet to the turbine. Currently systems are
being installed with throttle steam conditions as elevated as
1250–1550 psig/950°F. Such elevated conditions increase the
electricity/process heat ratio [21]. In addition to using
more severe throttle steam conditions, the FPI has improved
the performance of its boilers supplying steam to the cogeneration systems by adding economizers, closed shell-and-tube
feedwater heaters, and extended air heater surfaces that reduce the final boiler stack temperature to about 350°F.

In addition to the steam cycle improvements discussed above, considerable emphasis has been placed on the design of advanced burners such as the Lamb-Cargate wet cell or close-coupled updraft gasifier/combustor, suspension firing systems such as the Weyerhaeuser burner, and circulating fluidized beds such as those supplied by Ahlstrom Ltd. (Pyropower), Lurgi, and Foster-Wheeler. All such advanced burners achieve combustion while requiring relatively low levels of excess air (e.g., 15-25%) in order to improve the efficiency of wood fired systems.

Beyond the advances in the use of the steam cycle, the FPI is examining combustion turbine cogeneration systems to alter the electricity/process heat product ratio. Typically Brayton Cycle based cogeneration systems have electricity/process heat ratios that are up to double the power/heat ratios of comparable Rankine cycle systems [16]. Because electricity has a higher economic value than heat, such systems can be attractive.

A dramatic new development in wood energy is the fact that some utilities such as Washington Water Power, Spokane, Washington, and Burlington Electric, Burlington, Vermont, are using wood fuels in full condensing power stations. Private investors also are building such systems. Such conventional electricity generating units provide a market for wood fuels from both the mill and the forest, and investment opportunities for the FPI.

Until about 1973, the wood energy was the system required to consume low value residuals, and unify the integrated sawmill/pulp mill. Today the combination of PURPA markets and fuel sales has made energy a growth market for the FPI, as illustrated by Figs. 13 and 14. While the energy area may never rival sawmills or pulp mills in importance, it does provide another countercyclical area of capital investment.

D.   Product and Market Conclusions

The FPI produces a series of lumber and board products, pulp and related chemical products, and specialty products including energy. Of these lumber and pulp are most significant. In the area of lumber and board production, the industry is faced with severe overcapacity, and that condition is likely to extend into the foreseeable future. In the area of pulp production, the industry is facing a strong and growth oriented market at least in the short term. The industry also faces a significant cost squeeze in the area of pulp production that seriously threatens growth. Only in the areas of energy and chemicals does the industry face significant growth potential.

FIGURE 13. The 50 megawatt Joseph C. McNeill wood fired electricity generating station, Burlington, Vermont.

FIGURE 14. The 50 megawatt Brown Bovari turbine at the Joseph C. McNeill generating station, Burlington, Vermont.

III. CURRENT ECONOMIC AND FINANCIAL CONSIDERATIONS
     OF THE FOREST PRODUCTS INDUSTRY

     The economic organization and financial performance of
the FPI reflects the fact that this is a mature industry. It
produces auction type commodities. It emphasizes incremental
production cost improvements within existing product lines.
Analysis of potential responses by the FPI to the challenge
of change must consider these institutional issues.

A. *Structural Considerations of the FPI*

     The major research concerning the economic structure of
the FPI has been performed by Ellefson and coworkers [5, 9,
10, 26]. Additional research has been conducted by such
authors as Granskog [14], and Bethel *et al.* [4].

     1. Companies in the FPI. Today the FPI consists of
approximately 3,100 sawmills, 250 veneer mills, 310 plywood
mills, 140 particleboard mills, 300 pulp mills, and 740 paper
and paperboard mills [34]. The number of firms in each primary production category is a function of product demand plus
ease of entry into the marketplace. Entry costs can be considered as capital costs for new facilities, as shown in
Table III, plus additional administrative expenses.
     From Table III it is apparent that relative ease of entry
may characterize the lumber and board segment of the FPI, but
not the pulp and paper segment. This ease of entry issue
leads to considerations of market concentration. In lumber
and board manufacturing the top four producers command less
than 20% of the market and the top eight producers command
little more. In pulp manufacturing the top eight producers
command 83% of the market. The top eight paper producers
have garnered over 40% of the total market [17, 25].
     Pulp manufacturing, which is the most significant segment
of the FPI in terms of volume of product produced and revenue
generated, is the most difficult area to enter. Further,
pulp manufacturing is the most concentrated segment of the
industry. These two forces have led to a third critical
trend: production integration.

     2. Production Facility Integration. From a product manu-
facturing perspective, the most significant trend occurring
over the past 30 years has been the acquisition of pulp mills
by lumber and plywood manufacturers, and the acquisition of
lumber and board producers by pulp and paper companies. The
recently proposed merger of Champion and St. Regis highlights

TABLE III. Capital Cost of Major Forest Products Industry Facilities (1984 Dollars)

Facility type	Facility size	Facility cost
Sawmill	100×10^6 bd ft/yr	$ 27,700,000
Plywood mill	100×10^6 ft^2/yr (3/8" basis)	42,700,000
Particleboard mill	50×10^6 ft^2/yr (3/4" basis)	16,000,000
Pulp mill		
Unbleached kraft	1000 ton/day	358,500,000
Bleached kraft	1000 ton/day	414,500,000
Bleached sulfite	1000 ton/day	407,100,000
Thermomechanical	1000 ton/day	171,800,000
Paper mill		
Newsprint	1000 ton/day	254,000,000
Fine paper	1000 ton/day	280,100,000

Source: Tillman, Rossi, and Simmons [31] as updated.

this trend. By the mid 1970s Value Line Survey could no longer justify two categories of FPI companies for analysis, but switched to the classification "Paper and the Forest Products Industry."

The product integration trend within any given corporation (e.g., Weyerhaeuser) has extended horizontally to cover such commodities as lumber, plywood, engineered flakeboard, particleboard, mechanical pulp, chemical (e.g., kraft) pulp, paper products, energy products, and a host of other secondary goods. Product integration has occurred not only at the corporate level but also at the mill site level.

While horizontal (product) integration advanced, vertical integration also proceeded. Weyerhaeuser again provides a highly useful illustration as that firm: (1) owns over 5 million acres of forest land, and is therefore 100% self-sufficient in its basic raw material; (2) produces much of its own chlorine and caustic for pulping and bleaching; (3) is the fifth largest home builder in the U.S., providing an outlet for some of its lumber and board products while capturing more value added associated with solid wood products; and (4) is the second largest mortgage banker in the

U.S., providing financing for the ultimate products of its lumber and board manufacturing facilities.

The term integrated mills represents a general concept where a sawmill and possibly a plywood mill is physically located on the same site as a residue consuming facility such as a pulp mill or particleboard plant. Typical two dimensional studies concerning such mills focus on lumber and kraft pulp production as shown in Figs. 15 and 16 [35, 36]. Some studies include mechanical pulping in the models [35].

The construction of integrated mills achieves numerous process advantages including the following:

(1) log merchandising flexibility for optimum use of the timber as harvested in any given market (flexible standards can exist regarding the minimum size of logs entering the sawmill);

(2) maximum use of residues in materials production, without the expenditure of funds on nonproductive activities (e.g., pulp chip transportation);

(3) maximum energy conservation including the providing of heat sinks for waste heat generated by numerous processes (e.g., TMP pulping);

(4) maximum cogeneration potential by creating process energy demands sufficient to warrant more efficient boilers with elevated throttle steam conditions (see Kovacik [21] for a discussion of the relationship between heat sink size and cogeneration steam conditions);

(5) economies of scale in such capital investments as log handling systems, debarking systems, energy production systems, pollution control facilities, and general and administrative offices; and

(6) operating and maintenance gains from scale in such areas as plant maintenance and administration, product sales, and other general expenses.

The economic savings identified above are not trivial. In the energy area alone integration commonly permits cogeneration to be justified for the sawmill portion of the steam demand. The larger heat sink increases the cycle efficiency and consequent electrical generating capacity by as much as 20%. Waste heat recovery opportunities from TMP pulping and from other processes can exist due to integration, and they are equally beneficial.

One of the earliest integrated mills was assembled at Longview, Washington, by the Weyerhaeuser Company. Today

FIBER FLOWS IN THE POSTULATED INTERATED MILL
(BASIC 1 hr)

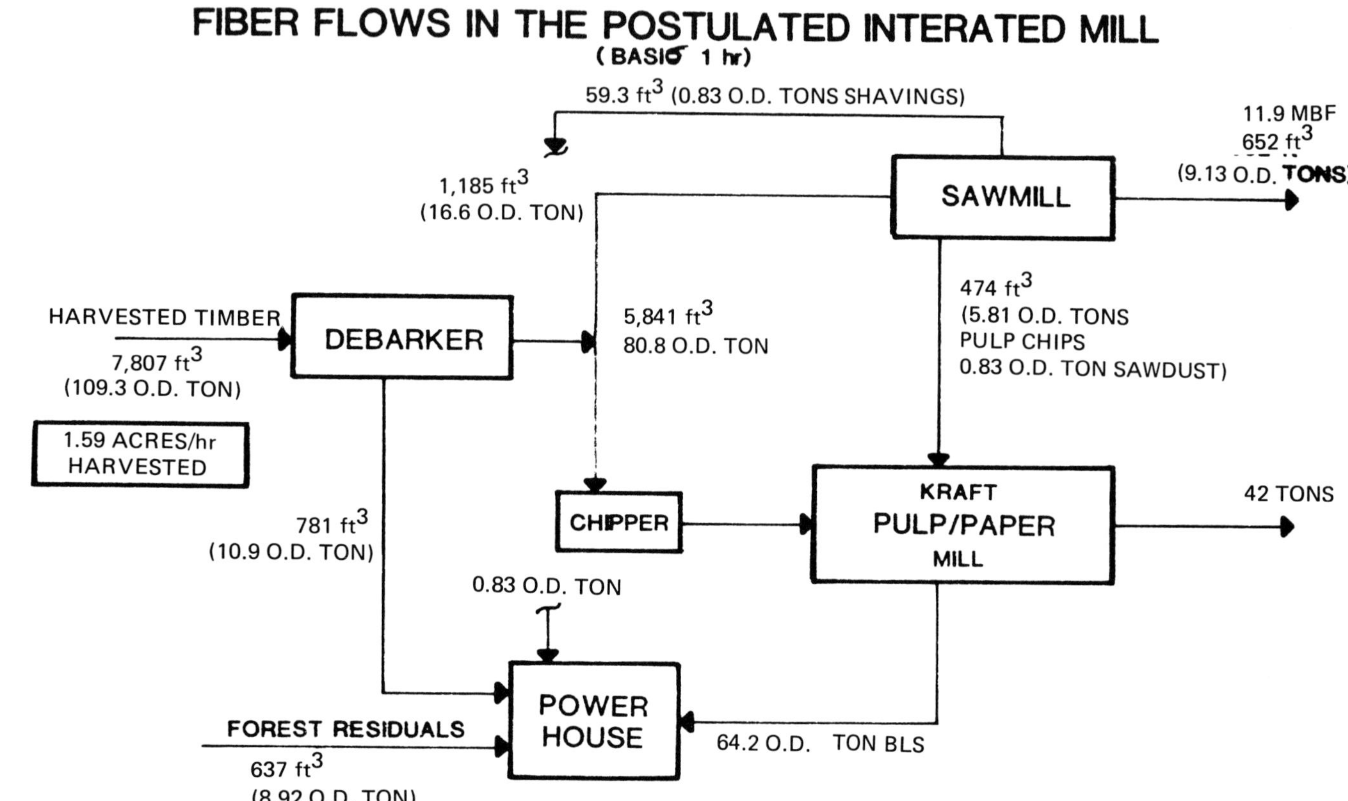

FIGURE 15. Fiber flows in a postulated 2-dimensional integrated mill. Source: [36].

STEAM AND POWER FLOWS IN INTEGRATED MILL

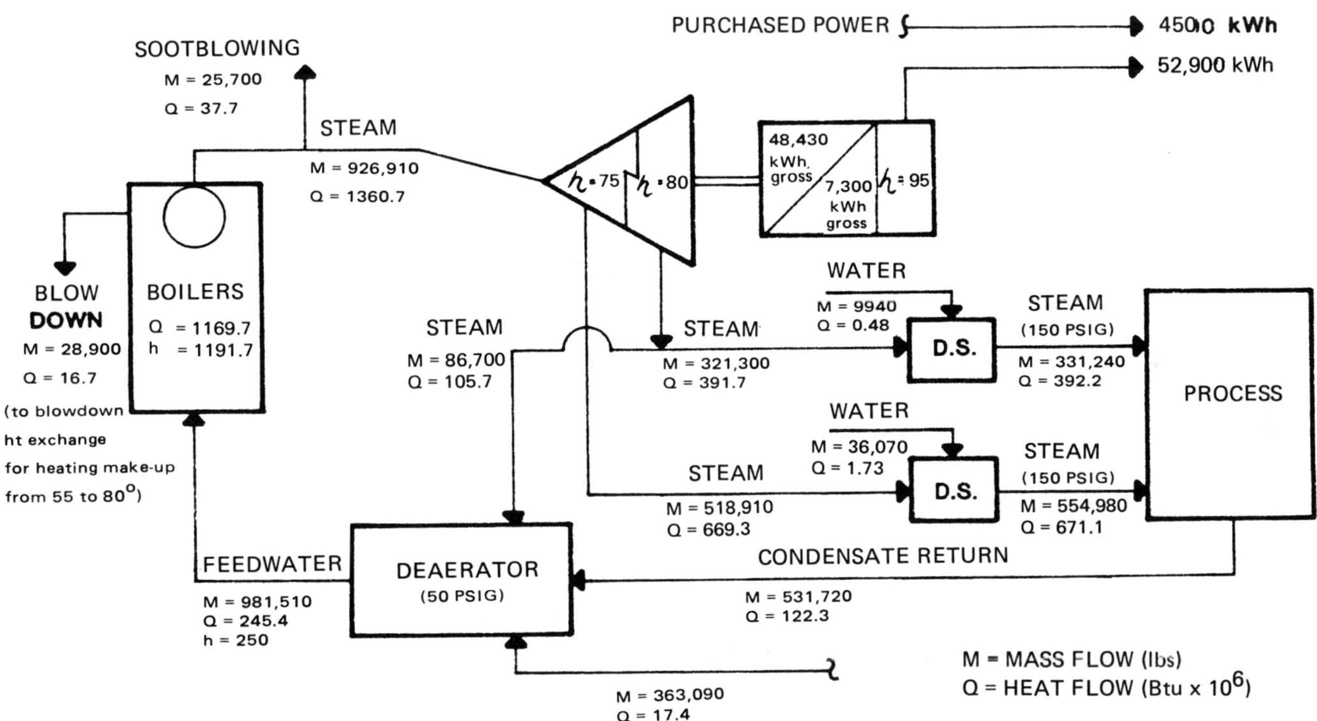

FIGURE 16. Energy flows for the postulated integrated mill shown in Fig. 15. Source [36].

27

that facility includes a large, modern sawmill, a plywood
mill, both kraft and TMP pulp mills, a chemical production
facility for making chlorine and caustic, and a 30 megawatt
cogeneration facility. Broad views of this facility are
shown in Figs. 17 and 18. Among the recent integrated mills to
be assembled is the Leaf River Forest Products, Inc. facility
owned by Great Northern Nekoosa and Kymi Kymmene Oy of Finland. The sawmill on that site has an annual capacity of
135×10^6 bd ft of lumber while the pulp mill is a 1000 ton/
day bleached kraft mill. The Leaf River facility also includes a 350,000 lb/hr power boiler and a 1500 ton/day
liquor recovery boiler, each producing 1250 psig/900°F steam.
Cogeneration is accomplished in two 25 MW turbogenerators
[27].

Other economically efficient integrated mills have been
built or assembled in such locations as Plymouth, North
Carlina, New Bern, North Carolina, and elsewhere.

Horizontal integration is of vital importance in that it
pushes the entire FPI production possibility frontier out-
ward. Further, it permits countercyclical revenue generation
due to the potential mix of products produced (e.g., lumber,
kraft pulp, electricity).

While integrated mills are highly efficient, they carry
the consequences of capital intensity, difficulty of market
entry, and economic concentration from the pulp mill portion
of the FPI to the entire FPI. Auction commodity producers
must be the lowest cost, most efficient manufacturers within
any industry in order to survive. Such efficiencies are
achieved by integration. While specialty products may be an
alternative route to success for some firms, such product
lines cannot be the basis for future growth of the FPI as a
whole.

B. The Financial Consequence of FPI Structure

The financial consequence of the current FPI structure
is a highly capital intensive, slow growing, industry. The
capital intensity can be shown by virtually any measure
chosen. Assets per employee, as reported by Fortune Magazine, is one appropriate measure. Table IV illustrates
assets per employee in the FPI, and in the average of all
manufacturing industry over time. It is useful to note that
the FPI is increasing its capital intensity relative to all
industry.

The FPI is highly capital intensive, ranking fifth among
the 27 manufacturing industry types in assets per employee.
Ranking ahead of the FPI, however, are its major competitors:

FIGURE 17. A broad overview of the Weyerhaeuser integrated mill, Longview, Washington. (Photo courtesy of Weyerhaeuser Co.)

FIGURE 18. A broad aerial view of the Longview complex, illustrating its size. (Photo courtesy of Weyerhaeuser Co.)

TABLE IV. Assets Per Employee in the Forest Products Industry, and in All Manufacturing Industries

	Assets per employee	
Year	*Forest products industry*	*All manufacturing industry*
1983	$109,000	$70,000
1980	76,000	56,000
1978	61,000	44,000
1977	57,000	40,000
1975	48,000	38,000
1974	42,000	34,000

Source: Fortune 500 Directories for 1984 [2]*, 1981, 1979, 1978, 1976, and 1975.*

petroleum refining, mining and crude oil production, chemicals, and metals manufacturing [2].

Capital intensity has caused an increase in the use of leverage in the FPI, as shown in Table V. Note that all of the firms shown in Table V have assets in excess of one billion dollars, and most have assets at least twice that large. Most of these firms have more than a billion dollars in stockholders equity as well. Despite such assets, growth has been financed increasingly by borrowing.

If the trends toward high capital cost and horizontal integration have increased the costs of entry and degree of concentration of the FPI, the capital structures of existing firms have accentuated the problem. The economic data, then, demonstrate that the issues of the challenge of change largely revolve around raising and allocating capital.

IV. SUMMARY AND PROSPECTUS.

The FPI faces numerous challenges as it approaches the rest of this century. Such challenges include developing a marketing orientation and strategy for growth, and investing in new processes and plants that will support such a program.

The lumber and board segment of the industry faces an uncertain and potentially shrinking market with tremendous overcapacity. The pulp and paper segment of the FPI faces

TABLE V.   Leverage in the Forest Products Industry

Company	Assets ($×10^6)	÷ Stockholders = Leverage[a] equity ($×10^6)	
Georgia Pacific	$4,979	$2,013	2.47
Weyerhaeuser	5,946	3,223	1.84
International Paper	5,617	3,321	1.69
Champion International	3,576	1,782	2.01
St. Regis Paper	2,880	1,531	1.88
Crown Zellerbach	2,411	1,135	2.12
Scott Paper	2,723	1,464	1.86
Union Camp	2,381	1,197	1.99
James River Corp.	1,152	344	3.35
Great Northern Nekoosa	1,818	902	2.02
Westvaco	1,488	825	1.80
Louisiana Pacific	1,265	729	1.74
Potlach	1,156	545	2.12

[a]*Leverage factor - 1 = debt/equity ratio.*
A leverage factor of 2 = a debt/equity ratio of 1.
A leverage factor of 1.5 = a debt/equity ratio of 0.5.

Source: Fortune, 1984 [2].

a growing market. At the same time pulp producers must confront the inevitable cost/price squeeze associated with intense competition. Confronting these problems means accelerating the rate of innovation within the FPI.

Energy, along with chemicals has become more of a distinct sector of the FPI, although a somewhat hidden one due to the internal sales and use of fuel, steam, and some electricity. Energy has become significant partly as a result of PURPA. It has become important, also, because of a fundamental restructuring in energy prices within the economy.

Economically the FPI has become increasingly concentrated, increasingly integrated, and increasingly capital intensive. These trends have resulted in increasing demands for capital within the FPI, and consequently increasing rates of leverage.

Like many mature industries, the FPI must attract new
capital and make technological investments essential for its
survival. Without such an infusion of capital and technology, this industry could continue to lose its competitive
position in the marketplace.

Opportunities do exist to improve the technology base,
hence economic performance, of the FPI. Those opportunities
exist within and outside the conventional product lines of
the FPI. Consequently the remainder of this text is devoted
to the analysis of such opportunities. It is therefore
organized to accomplish the following three tasks: (1) ex-
plaining and evaluating models appropriate for new investment
analysis; (2) positing discount rates appropriate for new
technology investment analysis within the FPI; and (3) evaluating some representative technologies in lumber and pulp
production, energy generation, and the production of specialty products.

This book presents opportunities in wood products manufacturing with some attention to the relative economic importance of various forest industry sectors. This relative
economic importance of various sectors is illustrated in
Table VI. Table VI presents representative revenues to the
FPI from its various primary products. The basic difference
between this table and most others of its type is that
Table V imparts an explicit value to wood fuels and energy.
Table V values wood fuel at its current marginal cost despite
the fact that over half of the wood fuel consumed is subjected only to transfer pricing. Industrial wood fuel consumed by the FPI and much household cordwood never enter the
commercial marketplace (see Lesnick [22] for a complete discussion of transfer pricing and other related phenomena
applied to biomass fuel).

On the basis of Table VI, the relative ranking of various
economic sectors within the FPI is as follows: (1) pulp,
(2) lumber, (3) energy (as fuel), (4) plywood, (5) chemicals,
and (6) composites (e.g., particleboard). If plywood and
composite products are aggregated together they do not have
the economic significance of wood fuels.

Table VI is illustrative of the direction of activity of
the forest products industry, as well as the hidden significance of wood energy and chemicals. At the same time too
much should not be made of the table due to its inherent
limitations, such as the following: (1) it uses representative production and price values from an economically unstable period; (2) it ignores many secondary products including paper, furniture, pallets, and the like; and (3) it fails
to disaggregate products by grade and consequent price (e.g.,
select structural vs shop lumber, OSB vs particleboard,
electricity vs hog fuel).

TABLE VI.   Representative Values for Various Wood Product Sectors

Sector	Annual production volume	Unit value (Typical, 1984$)	Total annual value ($×10^9$)	%
Lumber	32.3×10^9 BF	$250/MBF[a]	8.1	20.5
Plywood (ft^2 = 3/8" basis)	16.8×10^9 ft^2	$250/$ft^2 \times 10^3$	4.2[b]	10.6
Particleboard and and composites[b] (ft^2 = 3/4" basis)	3.9×10^9 ft^2	$100/MSF	0.4	1.0
Wood pulp (paper not incl.)			20.3	51.4
Bleached kraft	20.1×10^6 ton	$450/ton	(9.1)	
Unbleached kraft	19.6×10^6 ton	$350/ton	(6.9)	
Other	12.4×10^6 ton	$350/ton	(4.3)	
Energy (as fuel)	2.50×10^{15} Btu	$2/Btu \times 10^6$	5.0	12.7
Lumber, plywood, and board	271×10^{12} Btu		(0.5)	
Pulp and paper	1.07×10^{15} Btu		(2.1)	
Other industries	209×10^{12} Btu		(0.4)	
Household	950×10^{12} Btu		(1.9)	
Chemicals			1.5	3.8
Total			39.5	100.0

[a]This ranges from $150-300/MBF depending upon conditions. Current market prices are not used.
[b]If composites totally replaced plywood this sector would have an annual revenue value of 2.1×10^9/yr or 5.8% of the primary FPI revenue. Sources: [25, 31, 38].

    Despite these weaknesses, Table VI is useful for high-
lighting the declining role of plywood and panel products;
the dominance of pulping, and the relative rise in importance of wood fuels. This text, in particular Chapters V-VII,
attempts to reflect those relative rankings of products. In
doing so it places particular emphasis on analyses in central
and growth areas available to the FPI as it addresses the
challenge of change.

REFERENCES

1. Aho, W. O. 1983. Advances in Chemical Pulping Processes.
    In "Progress in Biomass Conversion," Vol. 4. Academic
    Press, New York. Also see Aho, W. O. 1984. Recent
    Developments in Chemical Pulping Technology. Presented
    at the 1984 Annual Meeting of the American Institute of
    Chemical Engineers, San Francisco, Nov. 25-28.

2. Anon. 1984. The 500: The Fortune Directory of the
    Largest U.S. Industrial Corporations. *Fortune*, Apr. 30,
    274-322. Also see *Fortune* 500 Directories for years
    1975-1983.

3. Bastia, N., Kummant, I., Schabas, W., Froelich, G.,
    Skole, R., and Tikkanen, T. 1984. New Pulping Methods—
    On Paper They Look Great. *Chemical Engineering* 91(10):
    22-26. For additional information on TMP and CTMP, see
    Cody, H. M. 1984. Quesnel River Pulp Converts TMP Mill
    to a 450-tpd CTMP Operation. *Pulp and Paper* 58(6):76-77.

4. Bethel, J. S. *et al.* 1976. Renewable Resources for
    Industrial Materials. National Research Council/
    National Academy of Sciences, Washington, D.C.

5. Bilek, E. M., and Ellefson, P. V. 1981. Wood-based
    Industry: Trends in Selected Structural and Economic
    Factors Through 1977. *Forest Products Journal* 31(10):
    48-54.

6. Boyd, J. *et al.* 1975. Problems and Legislative Oppor-
    tunities in the Basic Materials Industries. National
    Materials Advisory Board/National Academy of Sciences,
    Washington, D.C.

7. Diamond, S. 1984. Cogeneration Jars the Power Industry.
    New York Times Business Section. June 10, Sec. 33, pp. 1,
    28.

8. Ekono, Inc. 1983. Energy Use in California Pulp and Paper Industry. Ekono, Inc., Bellevue, Washington. For the California Energy Resources Conservation and Development Commission.

9. Ellefson, P. V., and Chopp, M. E. 1978. Systematic Analysis of the Economic Structure of the Wood-Based Industry. Institute of Agriculture, Forestry, and Home Economics, University of Minnesota, St. Paul, Minnesota.

10. Ellefson, P. V., and Chopp, M. E. 1979. U.S. Industrial Ownership of Forest Land: Extent, Strategies, and Management Intensity. Institute of Agriculture, Forestry, and Home Economics, University of Minnesota, St. Paul, Minnesota.

11. Enzer, H., Dupree, W., and Miller, S. 1975. Energy Perspectives. U.S. Department of the Interior, Washington, D.C.

12. Frederick, W. J. 1984. The Energy Costs of Increased Organics Recovery for Chemical Byproducts in Kraft Pulp Mills. *In* "Progress in Biomass Conversion," Vol. 5. Academic Press, New York.

13. Goetzel, A., and Tatum, S. 1983. Wood Energy Use in the Wood Products Industry: What the Data Show. *In* "Progress in Biomass Conversion," Vol. 4. Academic Press, New York.

14. Granskog, J. 1978. Economies of Scale and Trends in the Size of Southern Forest Industries. *In* "Complete Tree Utilization of Southern Pine," Proc. (C. McMillin, ed.). Forest Products Research Society, Madison, Wisconsin.

15. Grantham, J. B. 1974. Status of Timber Utilization on the Pacific Coast. U.S.D.A. Forest Service, Portland, Oregon.

16. Gyftopoulos, E. P., Lazaridis, L. L., and Widmer, T. F. 1974. "Potential Fuel Effectiveness in Industry." Ballinger Publishing Co., Cambridge, Massachusetts.

17. Hersh, H. N. 1981. Energy and Materials Flows in the Production of Pulp and Paper. Argonne National Laboratory, Chicago, Illinois.

18. Hurley, P. J. 1978. Comparison of Mill Energy Balance Effects of Conventional, Hydropyrolysis, and Dry Pyrolysis Recovery Systems. Institute of Paper Chemistry, Appleton, Wisconsin.

19. Koenigshof, G. A. 1979. Status of Com-Ply Floor Joist Research. *Forest Products Journal* 29(1):37-42.

20. Koning, J. W. 1984. Resource Impacts of New Developments in Pulp and Paper Manufacturing. Presented at the 38th Annual Meeting, Forest Products Research Society, St. Louis, Missouri, June 27. Also see Ince, P. J., Economics of Utilizing Hardwoods for Kraft Linerboard Production. Presented at the 38th Annual Meeting, Forest Products Research Society, St. Louis, Missouri, June 28.

21. Kovacik, J. M. 1980. Design Considerations for Large Industrial Cogeneration Systems. Presented at the Oregon Governor's Cogeneration Workshop, Portland, Oregon, April 20-21.

22. Lesnick, E. 1984. Microeconomic Approaches to Biomass Fuel Pricing. *In* "Progress in Biomass Conversion," Vol. 5. Academic Press, New York.

23. Libby, C. E., ed. 1962. "Pulp and Paper Science and Technology: Vol. 1, Pulp." McGraw-Hill, New York.

24. Maloney, T. C. 1977. "Modern Particleboard and Dry Process Fiberboard Manufacturing" Miller Freeman, San Francisco.

25. National Forest Products Association. 1981. Economic Commentary: The Wood Products Industry—A Brief Profile. NFPA, Washington, D.C.

26. O'Laughlin, J., and Ellefson, P. V. 1982. New Diversified Entrants Among U.S. Wood-Based Companies: A Study of Economic Structure and Corporate Strategy. Agricultural Experiment Station, University of Minnesota, St. Paul, Minnesota.

27. Patrick, K. 1984. Leaf River: Showcase Market Pulp Mill Makes Its Debut in Mississippi. *Pulp and Paper 58*(9): 60-72.

28. Rickford, E. 1984. Comments on Bucking and Merchandizing. Presented at the 38th Annual Forest Products Research Society Meeting, St. Louis, Missouri, June 27.

29. Risbrudt, C. D., and Stone, R. N. 1980. Trends and Patterns in Wood Products Consumption and Production. *In* "Timber Demand: The Future Is Now," Proc. Forest Products Research Society, Madison, Wisconsin.

30. Spelter, H., and Phelps, R. B. 1984. Changes in Postwar U.S. Lumber Consumption Patterns. *Forest Products Journal 34*(2):35-41.

31. Tillman, D. A., Rossi, A. J., and Kitto, W. D. 1981. "Wood Combustion: Principles, Processes, and Economics." Academic Press, New York.

32. Tillman, D. A., Schnorr, R., and Sale, J. 1982. Biomass as a Fuel: An Incremental Analysis of Heat Rates. *In* "Progress in Biomass Conversion," Vol. 3. Academic Press, New York.

33. Tillman, D. A., and Jamison, R. L. 1982. Cogeneration With Wood Fuels: A Review. *Fuel Processing Technology* 5:169–181.

34. Tillman, D. A., Rossi, A. J., and Simmons, S. O. 1982. Wood: Its Present and Potential Uses. Envirosphere Co., Bellevue, Washington. For the U.S. Congress, Office of Technology Assessment.

35. Tillman, D. A. 1983. Evaluating the Potential Contribution of the Forest Products Industry to U.S. Energy Supply. *Journal of Applied Polymer Science, Applied Polymer Symposium* 37:593–615.

36. Tillman, D. A. 1983. Making the Best Energy Use of Wood. *In* "Progress in Biomass Conversion," Vol. 4. Academic Press, New York.

37. Ulrich, A. H. 1981. U.S. Timber Production, Trade, Consumption, and Price Statistics 1950–1980. U.S.D.A. Forest Service, Washington, D.C.

38. U.S. Department of Commerce. 1984. Statistical Abstract. U.S. Government Printing Office, Washington, D.C.

39. Walker, J. E. 1966. "Hopewell Village: The Dynamics of a Nineteenth Century Iron-Making Community." University of Pennsylvania Press, Philadelphia.

40. Wolpert, V. M. 1977. "Synthetic Polymers and the Paper Industry." Miller Freeman, San Francisco.

Chapter II

ANALYSIS OF FOREST INDUSTRY INVESTMENTS:
THE BASIC CASH FLOW MODEL

I. INTRODUCTION

The Forest Products Industry faces unprecedented challenges in converting from a resource based industry to a market based industry, and from a natural growth-large log resource to a managed growth-small log resource. This challenge must be met with new and vastly improved processes in those mills converting timber into the myriad of forest products. The problem for engineers and wood technologists, then, is the evaluation of candidate processes for change. Addressing that problem requires the use of analytical models that facilitate accurate process evaluation. Further, it requires tailoring those models to the specific needs of new technology evaluation.

Several models can be used in process analysis: (1) the discounted cash flow (DCF) model, (2) comparative economic models derived from the DCF model, and (3) comparative physical models. Of these, the DCF model and its derivatives provide decision making values. Because the DCF model provides decision oriented values, it is the primary tool used here.

This chapter reviews the DCF model. It examines how this model can be used to accomplish a variety of tasks from investment analysis to investment comparison and product pricing. It is sufficiently flexible that it can handle problems of inflation, project financing, product pricing, and high risk situations.

II. THE DISCOUNTED CASH FLOW APPROACH FOR EVALUATING INVESTMENTS

Efficiency is one of the critical concepts for the industry as it faces the coming decades. Efficiency here does not mean only measuring the percentage yield of lumber or pulp, or percentage of energy self-sufficiency in a mill. Efficiency here implies the ability of processes and products

to generate cash flows for forest industry corporations. This
is efficiency in engineering economy.

Several models can be used to measure the economic efficiency as the absolute or relative desirability of a given
process. These are the standard DCF, levelized cost, and
strategic investment models. All of these models are based
upon the present worth concept shown in the following formula:

$$V_a = V_0 (1 + i)^{(y_a - y_0)} \qquad\qquad (2\text{-}1)$$

Where V_a is the value of some sum of money in future year a,
V_0 represents some dollar value at the present time (time 0),
i is some interest rate, y_a is that project year in the
future when V_a is calculated, and y_0 is the current project
year. For convenience, the exponent can be expressed as
follows:

$$y_a - y_0 = t \qquad\qquad (2\text{-}2)$$

Where t is the time period, in years, used in the analysis.
The correlary formula to (2-1) is as follows:

$$V_0 = \frac{V_a}{(1 + i)^t} \qquad\qquad (2\text{-}3)$$

A series of additional expressions defining present values of
annuities, sinking fund requirements, capital recovery factors and future values of annuities can be derived from these
equations. The details of these derived expressions are described elsewhere (see, for example, Neuman [12], Schall and
Haley [14], Bierman and Smidt [1]). What is important here
is the use of the basic concept, time value of money, in
project assessment. This concept is the foundation for the
DCF models used in project analysis.

A. Discounted Cash Flow Mathematics

The basic DCF model, as discussed in engineering economy
and corporate finance texts, provides corporations with certain absolute measures of whether or not to fund projects.
Those measures are the Net Present Value (NPV) and Internal
Rate of Return (IRR). Those basic measures lead to deriva-
tive values such as levelized cost.

1. The Basic Equations of the Cash Flow Model. NPV is
a measure of the increase in the present value of a corporation's wealth that can be achieved by undertaking a given

project. Mathematically it is calculated as follows:

$$\text{NPV} = \sum_{y=0}^{a} \frac{\text{NATCF}_y}{(1+dr)^y} - (I - \text{ITC}) \qquad (2\text{-}4)$$

Where y is any given project year from the present (y_0) to the terminal year of the project (year a). The critical variables are NATCF, which represents the net after tax cash flows during the operation of the project; dr, which represents some rate of discount; I, which represents the investment; and ITC, which represents the investment tax credits.

NATCF$_y$ for any year represents the cash generated by a project in that year, and is calculated as follows:

$$\text{NATCF}_y = (R_y - \text{O\&M}_y - D_y) \times (1 - \text{TR}) + D_y \qquad (2\text{-}5)$$

Where R is the revenues generated by the project, O&M represents the fixed and variable operation and maintenance costs, D is the depreciation taken in that year, and TR is the tax rate (expressed as a decimal).

IRR is a measure of the percentage return achieved by a project, and it is calculated with the same variables used to calculate NPV, solving the following expression for r:

$$I - \text{ITC} = \sum_{y=0}^{a} \frac{\text{NATCF}_y}{(1+r)^y} \qquad (2\text{-}6)$$

Alternatively,

$$\sum_{y=0}^{a} \frac{\text{NATCF}_y}{(1+r)^y} - (I - \text{ITC}) = 0 \qquad (2\text{-}7)$$

Essentially solves for the rate of discount when NPV = 0.

Using equations (2-4) through (2-7) can be illustrated by an example, which starts simply and then is modified to accommodate increasing complexities in analysis (e.g., the handling of inflation). Such an example depends upon constructing a pro forma statement. The pro forma illustrates the use of depreciation and loss carry forward as cash flow generators in the calculation of the basic measures NPV and IRR.

The example employed is a $1 million investment, generating $500,000/yr in revenues over its 10 year life. It is depreciated over 5 years according to the schedule shown in Table I. It has operating and maintenance costs of $100,000/yr. The investing company has an income tax rate of 46%, and has an after tax discount rate (cost of capital) of 8%.

*TABLE I.   Depreciation Schedules for Projects Placed
on Line After 1985 (Accelerated Cost Recovery System)*

Year of service	Percentage depreciation			
	3-year schedule	*5-year schedule*[a]	*10-year schedule*	*15-year schedule*[b]
1	33	20	10	7
2	45	32	18	12
3	22	24	16	12
4		16	14	11
5		8	12	10
6			10	9
7			8	8
8			6	7
9			4	6
10			2	5
11				4
12				3
13				3
14				2
15				1

[a]*Typically this would be used for sawmills and pulp mills cogeneration facilities owned by the FPI.*

[b]*Typically this would be used for public utility investments.*

Source: Public Law 97-34 [13].

The pro forma statement for this example is shown in Table II. The NPV is $927,490, and the IRR is 26.4%.

2. Inflation and the Cash Flow Model. The DCF model does not address the problem of inflation, the general rise in prices over time, directly. Inflation began presenting serious problems in engineering economy analysis after about 1967. It introduced additional uncertainties into projecting cash flows. Incorporation of inflation into the analysis

TABLE II.   Pro-Forma Statement for Example of NPV and IRR Calculation
(Values in thousands of dollars)

Income/Expense		Year										
		0	1	2	3	4	5	6	7	8	9	10
Investment		(1000)										
Revenue			500	500	500	500	500	500	500	500	500	500
O & M costs			100	100	100	100	100	100	100	100	100	100
Depreciation			200	320	240	160	80	–	–	–	–	–
Earnings before taxes			200	80	160	240	320	400	400	400	400	400
Taxes			92	37	74	110	147	184	184	184	184	184
Net income after taxes			108	43	86	130	173	216	216	216	216	216
Investment tax credit		100										
Depreciation			200	320	240	160	80	–	–	–	–	–
Net after tax cash flow		(800)	308	363	326	290	253	216	216	216	216	216

Net Present Value @ 8% discount rate = 927.49

Internal Rate of Return   =   26.4%

requires measuring general and specific rates of inflation or
price escalation, and developing a methodology for treating
those rates in the analysis.

Inflation can be considered to be the general erosion in
the value of a dollar over time, as well as the general rise
in prices. Inflation has reached levels exceeding 10%/yr
within the past decade. It has created problems with establishing the economic value of assets in place. It has eroded
the value of depreciation as a cash flow generator. It has
raised the nominal discount rates associated with manufacturing industry. It has raised real discount rates by increasing business uncertainty. The seriousness of inflation as it
impacts accounting procedures and the construction of pro
forma statements should not be understated. Various impacts
of inflation are discussed by Seed [15] and Todd [18].

Inflation is measured by the Gross National Product
Implicit Price Deflator Series, the Consumer Price Index, and
the Producer Price Index. The Implicit Price Deflator Series,
shown in Table III, is generally regarded as the best indi-
cator of general inflation because it reflects price changes
of all goods and services. Over the long term, the three
measures move in the same direction and by about the same

TABLE III. *The Inflation Rate as Measured by the
Implicit Price Deflator Series*

Year	Percentage inflation
1973	5.8
1974	9.7
1975	9.6
1976	5.3
1977	5.5
1978	7.3
1979	8.6
1980	9.3
1981	9.4
1982	6.0
1983[a]	4.5
1984[a]	3.9

Source: U. S. Department of Commerce [19]. [a]*Estimated.*

amount, but the magnitude of change can vary markedly over
short periods of time. Inflation, as measured by the GNP
deflator series, exceeded 9% in 1980 and 1981, and then declined to 6% in 1982. It then declined to less than 4%
in 1983 and 1984. Specific goods and services exhibit escalation rates that may be above or below the general rate of
inflation. Some comparative rates are shown in Table IV.
Investments in new plant and equipment also may escalate at
rates slower or faster than inflation.

Inflation influences all projected revenue and cost
streams, and the discount rate. Therefore it must be treated
explicitly in the construction of pro forma cash flow statements. Three approaches are used to account for the effects
of inflation:

(1) calculate the pro forma cash flows based upon
    "nominal" dollars,

(2) calculate the pro forma cash flows based upon "real"
    dollars, and

(3) calculate the pro forma cash flows based upon
    "constant" dollars.

Nominal dollars incorporate inflation into all variables,
including the discount rate. Real dollars explicitly remove
inflation from the cash flows and the discount rate. Constant dollars assume that inflation does not exist. Table II
used a constant dollar approach. For purposes of highlighting the difference between analytical techniques, it has been
reconstructed in Table V assuming a rate of general inflation of 7%, applied to revenues and operating and maintenance costs. The resulting measures of economic desirability
are shown in Table VI.

Table V is worth examining and comparing to Table II. Of
signifiance is the fact that the inflation free cash flows
for years 1-5 are less in Table V than in Table II. This is
caused by the fact that depreciation and taxes are calculated on a nominal dollar basis for tax purposes. Thus the
constant dollar approach always overstates the cash flows
generated by depreciation, and therefore gives the most opti-
mistic results. Only after depreciation is no longer in
effect are the real and constant dollar cash flows equivalent.

The comparative consequences of the cash flow analyses
are shown in Table VI. Note that the deflator series for
Table V has been calculated by the expression:

$$\text{def}_y = (1 + \hat{\imath})^y \tag{2-8}$$

TABLE IV.   *Escalation Indices for Various Commodities*

	Commodity						
Year	Producer price index	Pulp, paper and allied products	Lumber and wood products	Processed foods	Farm products	Fuels and energy	Chemical and allied products
1965	96.6	96.2	95.9	95.5	98.7	95.5	99.0
1970	110.4	108.2	113.6	112.1	111.0	106.2	102.2
1975	174.9	170.4	176.9	182.6	186.7	245.1	181.3
1979	235.6	219.0	300.4	222.5	241.4	408.1	222.3
1980	268.6	249.3	288.8	241.0	249.3	573.4	260.3
1981	294.8	272.9	298.1	241.2	249.4	707.6	274.3
1982	299.4	289.3	288.7	255.8	252.7	677.4	293.5
Long term escalation rate (%)	6.9	6.7	6.7	6.0	5.7	12.2	6.6

Source: U. S. Department of Commerce [19].

TABLE V.   *Nominal and Real Cash Flows Calculated from the Example in Table II.*
(Values in thousands of dollars)

Income/Expense		Year										
	0	1	2	3	4	5	6	7	8	9	10	
Investment	(1000)											
Revenue		535	572	613	655	701	750	803	859	919	983	
O & M cost		107	114	123	131	140	150	161	172	184	197	
Depreciation		200	320	240	160	80	–	–	–	–	–	
Earnings before taxes		228	138	250	364	481	600	642	687	735	786	
Taxes		105	63	115	167	221	276	295	316	338	362	
Net income after taxes		123	75	135	197	260	324	347	371	397	424	
Investment tax credit	100											
Depreciation		200	320	240	160	80	–	–	–	–	–	
Nominal net after tax cash flows	(900)	323	395	375	357	340	324	347	371	397	424	
Deflator	1.00	1.07	1.14	1.23	1.31	1.40	1.50	1.61	1.72	1.84	1.97	
Real after tax cash flows	(900)	302	346	305	272	243	216	216	216	216	215	

TABLE VI.   Measures of Financial Desirability for the Example in Tables II and V.

Dollar base	Discount rate (%)	Net Present Value	Internal Rate of Return
Nominal	16^a	$840,720	38.0
Real	8	870,190	29.0
Constant	8	927,490	30.5

[a] $1.07 \times 1.08 = 1.16$

Where def = deflator and i = the expected rate of inflation (decimal basis). The conversion from nominal to real discount rate for Table VI is calculated by the following expression [4]:

$$dr_r = \left(\frac{1+dr_n}{1+\hat{i}}\right) - 1 \tag{2-9}$$

Where dr_r = real discount rate and dr_n = nominal discount rate. The overstatement effect of the constant dollar approach is shown in Table VI, where both the NPV and IRR are higher than the NPV and IRR as calculated by the real dollar approach. In this case the results are positive for both analyses. However if the dr_r used for NPV calculation were set at 30%, the real dollar analysis would yield a negative NPV and the investment would not be made. The constant dollar analysis, however, would offer a NPV = $11,470 and the investment would be favored. The difference between go and no-go, as presented above, results from overstating the cash flows generated by depreciation.

NPV analysis is largely indifferent to whether nominal or real dollars are used. Businesses operate in a nominal dollar world. However, it is sometimes easier to understand the real dollar analysis because the dollar basis is intuitively familiar to the reader. Both bases are employed in this text. Nominal dollars are favored for final values in the evaluations within this text unless otherwise noted because those dollars do occur. Constant dollars are inaccurate and not used.

3. Debt, Equity and the DCF Model. To this point it has been assumed that the discount rate includes the cost of equity and the after-tax cost of debt such that the tax implications of borrowing have been incorporated into the analysis. This approach is discussed in Schall and Haley [14] Haley and Schall [7], and other texts. It provides for the separation of "what to finance" decisions from "how to finance" decisions, while avoiding the double counting of the interest deduction. It includes the implicit assumption that corporations do not change capital structures for individual projects. Models exist that accommodate the potential for borrowing various portions of capital, however, and they merit consideration here.

If borrowing is to be considered, then equations (2-4) and (2-5) must be adjusted as follows:

$$NPV_e = \sum_{y=o}^{a} \frac{NATCF'_y}{(1+k_e)^y} - (I_e - ITC) \tag{2-10}$$

$$NATCF'_y = (R_y - O\&M_y - INT_y - D_y) \times (1 - TR) - PP_y + D_y \tag{2-11}$$

Where NPV_e is the net present value of the equity investment; NATCF' is the net after tax cash flow, incorporating the tax deductibility of interest expense on borrowed money; k_e is the cost of equity, or the equity portion of the discount rate; I_e is the equity funded portion of the investment, INT is the annual interest payment, and PP is the principal payment. IRR can be recalculated by adjusting equation (2-6) as follows, and solving for r':

$$I_e - ITC = \sum_{y=o}^{a} \frac{NATCF'_y}{(1+r')^y} \tag{2-12}$$

The pro forma shown in Table V is recalculated to illustrate this approach. Prior to use of this pro forma, however, it is necessary to estimate k_e, as the problem no longer assumes an overall discount rate. If one treats dr by the weighted average cost of capital (WACC) technique, then the expression is [7]:

$$dr = \theta k_d + (1 - \theta) k_e \tag{2-13}$$

Where θ is the proportion of debt and k_d is the after tax cost of debt. The value k_d must be converted to k_{dp}, or the pretax cost of debt (the interest rate charged by the bank or bond holder) as follows:

$$k_{d_p} = \frac{k_d}{(1-TR)} \qquad\qquad (2\text{-}14)$$

The example shown in Table V now can be recalculated using a debt/equity ratio of 1 ($\theta = 0.5$), and a $k_{d_p} = 14\%$ (nominal). On this basis $k_d = 7.6\%$, and $k_e = 24.4\%$. Note that the expected return on equity is substantially higher than the expected return on debt. This is due to the inherently higher risk associated with equity investments.

The cash flows as adjusted are shown in Table VII. The resulting financial measures are as follows: (1) NPV_e (nominal) = \$650,560 ($DR_e = 24.4\%$); (2) $ROE_n = 71.3\%$; (3) NPV_e (real) = \$649,300 ($DR_e = 16.3\%$); and (4) $ROE_r = 60.1\%$. ROE is used to denote return on equity either on a nominal (n) or a real (r) basis.

There is an intuitive appeal to this debt/equity model in that it replicates, in more detail, the real accounting world. The decision results of the debt/equity model have been compared to the decision results of the previous model shown in equation (2-4) through (2-6) and Table V, however; and in no case has the go/no-go decision been reversed. The difference is in the handling of the tax treatment of interest payments. This debt/equity model is not used in subsequent analysis here, however, because it assumes that the engineer or wood technologist, rather than the chief financial officer of a company, will determine the project financing, and the impact of a project's financial structure on the firm as a whole. The simpler model duplicates the division of responsibilities in a major corporation by holding the corporation's financial structure constant.

B. Uses of the DCF Model

The DCF model defined in equations (2-4) through (2-7) and modified by equations (2-8) and (2-9), is the primary decisionmaking tool for evaluating new capital projects on a go/no-go basis. On that basis it assumes that the plant is a "price-taker," that the sales value of the commodity is determined by forces of supply and demand rather than technology. For new technologies, the minimum required price of commodities to be produced also can be derived from the DCF model. This application, along with decisionmaking, is discussed below.

1. Using the DCF Model for Decisionmaking. The NPV and IRR, as calculated in equations (2-4) through (2-9), are

TABLE VII.   Nominal and Real Cash Flows With Accounting for Debt and Equity Based Upon Table V.

Income/Expense		Year										
	0	1	2	3	4	5	6	7	8	9	10	
Investment	(1000)											
Revenue		535	572	613	655	701	750	803	859	919	983	
O & M costs		107	114	123	131	140	150	161	172	184	197	
Depreciation		200	320	240	160	80	–	–	–	–	–	
Interest expense		70	66	62	57	49	43	35	27	17	6	
Earnings before taxes		158	72	188	307	432	557	607	660	718	780	
Taxes		73	33	86	141	199	256	279	304	330	359	
Net income after taxes		85	39	102	166	233	301	328	356	388	421	
Proceeds from borrowing	500											
Principal repayment		26	30	34	39	47	53	61	69	79	44	
Investment tax credit	100											
Depreciation		200	320	240	160	80	–	–	–	–	–	
Net after tax cash flow	(400)	259	329	308	287	266	248	267	287	309	377	
Deflator	1.00	1.07	1.14	1.23	1.31	1.40	1.50	1.61	1.72	1.84	1.97	
Real after tax cash flow	(400)	242	289	250	219	190	165	166	167	168	191	

measures used in investment decision making. For any given
discount rate or hurdle rate (weighted average cost of capital), a corporation will make a go/no-go decision if (1) the
NPV is a positive number, or (2) if the calculated IRR
exceeds the hurdle rate. Further, under capital constraints,
the firm can rank investments by NPV or IRR, funding those
with the highest calculated NPV or IRR value first, and funding subsequent projects until the capital budget is fully
allocated.

There is considerable debate in the literature concerning
the relative desirability of the NPV and the IRR, and conventional wisdom favors the NPV. The arguments favoring the NPV
are as follows:

(1) The use of the NPV maximizes the wealth of the corp-
    oration. The use of the IRR focuses on growth rather
    than wealth.

(2) The IRR is subjected to certain mathematical vaguer-
    ies. If the NATCF term calculation in equation (2-5)
    temporarily changes sign (goes from positive to nega-
    tive) in the years after the investment, multiple
    mathematically accurate IRR values can be obtained.
    The selection of the appropriate IRR value among mul-
    tiple solution becomes subjective.

(3) The IRR contains a reinvestment assumption, that the
    cash flows generated by the project are reinvested at
    the calculated rate of return. The NPV also allegedly
    has a reinvestment assumption, with the reinvestment
    being made at the discount rate. The NPV reinvestment
    assumption is considered more appropriate [3].

The debate on the relative desirability of the NPV and
IRR measures is far from resolved. The debate has several
sub-issues including the very existence of the reinvestment
assumptions [2]. It is neither possible nor useful to attempt
to resolve these debates here. In general this text employs
the NPV and shows the IRR for illustrative purposes only. The
reason for using the NPV as the primary decision making criteria is that the Forest Products Industry is led by large,
mature firms. It does not operate in a high growth mode. Its
investments in new technology are designed to protect and
enhance the wealth of the stockholders.

Using the NPV in preference to the IRR typically does not
change the investment decisions made by a company. However,
under certain circumstances, mutually exclusive investments
may be ranked differently. The NPV measure is influenced by
the size of a prospective investment. The IRR measure is not

influenced by the size of the capital investment opportunity
available. This use of NPV is consistent with the growth in
forest industry facility sizes (see Granskog [5]).

For these purposes, then, a favorable NPV at a calculated
discount rate will be used to demonstrate potentially attractive investments.[1] Such a focus in the decision making process restricts the issues of interest in assessing new technologies to the following: (1) revenue estimates, (2) the
capital and operating and maintenance cost estimates, (3) the
appropriate discount rate to be used, and (4) techniques for
handling unusual business risk. These issues subsume ques-
tions concerning process yield, energy efficiency, and envir-
onmental consequences; and convert those issues into dollar
values.

2. *The Use of the DCF Model for Balance Sheet Analysis.*
The DCF model can evaluate the impact of a new investment on
the balance sheet of a company. This is an extension of the
decision analysis use of DCF. The DCF model uses one of the
basic accounting tools, the income statement, as a basis for
calculation. The pro forma cash flow statement shown in
Table II is an income statement with modifications to the
treatment of depreciation, and with the elimination of an
interest expense line. Table VII is even closer to the traditional income statement, as the interest line is included.

The DCF model can relate the income statement to the balance sheet, illustrating the impact of a project on the
future health of the corporation. To do this it must be
recognized that the balance sheet is based upon a simple
equation:

$$A = L + OE \tag{2-15}$$

Where A = assets, L = liabilities, and OE = owners equity.
The retained equity portion of cash flows becomes an addition
to owners equity. This is illustrated in Table VII, where
debt holders have been repaid and the remaining cash flow
belongs to the owners. These remaining cash flows either are
paid to the owners as dividends, or are reinvested in the
firm (e.g., as owners equity).

---

[1] *Constraints of interrelated investments, investments
with differing economic lives, and selection of projects
under severe capital budgeting decisions also influence the
use of this measure. They are treated in numerous texts
(e.g., Bierman and Smidt [1]), and are outside the purpose
of this analyses.*

To relate the pro forma cash flow statement to the balance sheet requires the rewriting of equation (2-4) or (2-10) as follows:

$$NFV = \left(\sum_{y=o}^{a} NATCF_y (1 + dr)^y \right) - (I - ITC) \times (1 + dr)^t$$

$$(2\text{-}16)$$

Where NFV is the net future value of a project. If (2-10) is rewritten, k_e is substituted for dr, NFV_e is substituted for NFV, $NATCF'$ for NATCF, and I_e for I.

The NFV has an explicit reinvestment assumption, that the cash flows are reinvested at the discount rate or the cost of equity, depending upon the approach used. The NFV can be calculated either on a nominal or a real basis.

The relationship between the DCF model and the balance sheet comes from the use of the NFV or NFV_e term. Two more equations are required to show this approximate relationship:

$$NFV (1 - \theta) \cong \Delta OE \qquad (2\text{-}17)$$

$$NFV_e = \Delta OE \qquad (2\text{-}18)$$

Where ΔOE = the change in owners equity. The NFV approach measures the increase in assets that could be supported by equity investments as a result of any investment assuming no change in liabilities, and given the equation:

$$\Delta A = \Delta L + \Delta OE \qquad (2\text{-}19)$$

Where ΔA is the change in assets and ΔL is the change in liabilities. Alternatively, if one assumes no outside change in corporate debt or cost of capital, one can then assume:

$$NFV_e \cong \Delta A \qquad (2\text{-}20)$$

The NFV, then, can be used to estimate the balance sheet consequences of a project at the end of its investment life. If NFV > 0, then the investment will promote asset growth of a corporation. If NFV < 0, the project will lead to a reduction in the asset base of a corporation. Thus, through the use of the DCF model, the impact of a project can be measured with respect to corporate profits, taxes, cash flows, owners' equity, and assets.

3. *The Use of the DCF Model in Product Pricing.* The final use of this model is in establishing minimum acceptable product prices for goods produced by new technologies, and the consequent gains that can be made by bringing a new

technology to maturity. This application is particularly
useful in evaluating whether new technologies can compete in
the marketplace when introduced under unusual risk conditions, and their market potential under mature technology
conditions.

Mathematically this use of the model depends upon establishing the capital recovery factor by the following equation [12]:

$$CRF = \frac{dr(1+dr)^t}{(1+dr)^t - 1} \qquad (2-21)$$

Where CRF = the capital recovery factor at some after-tax discount rate.

The annuity type price is then calculated by the following expressions:

$$NATCF_{req} = (I - ITC)\ CRF \qquad (2-22)$$

$$\frac{(NATCF_{req} - D)}{(1-TR)} + O\&M + D = RR \qquad (2-23)$$

$$\frac{RR}{Q} = MSP \qquad (2-24)$$

Where RR is annual revenue requirements in a given year, Q is
the annual quantity of product produced in that year, and MSP
is the minimum sales price in \$/unit in the year under con-
sideration. It is most convenient to calculate prices originally on a nominal dollar basis. Prices can then be converted to a real dollar basis through use of the appropriate
deflator prior to their display.

The use of this model can be illustrated by the example
previously shown in Table V. To do so requires only one
additional assumption: that the system produces one million
units of product per year. The example is shown in Table
VIII.

Note that losses in the early years, shown in Table VIII,
are carried forward to later years as a noncash charge, or
tax avoidance measure. They are treated in an accounting
manner not unlike depreciation. Note, also, that there is an
uneven price depending upon the influences of depreciation,
tax losses, and escalation. These prices can be used in the
comparative analysis. The price range, however, is typically
relatively narrow when established on a real dollar basis.
These prices can be levelized for the 10 year time period,
as is shown in the following expression:

TABLE VIII. *Minimum Price Requirements for the Production of One Million Units of Product Based on Table V.* *(Values in thousands of dollars)*

Parameter		Year										
	0	1	2	3	4	5	6	7	8	9	10
NATCF required (dr = 16%)		186	186	186	186	186	186	186	186	186	186
Depreciation		200	320	240	160	80	–	–	–	–	–
Loss carried forward					26	106	70				
Investment tax credit	100										
Net income after taxes		(14)	(134)	(54)	–	–	116	186	186	186	186
Taxes		–	–	–	–	–	99	158	158	158	158
Earnings before taxes		(14)	(134)	(54)	–	–	215	344	344	344	344
Loss carried forward					26	106	70				
Depreciation		200	320	240	160	80	–	–	–	–	–
O & M cost		107	114	123	131	140	150	161	172	184	197
Revenue		293	300	309	317	326	435	505	516	528	541
Investment	(1000)										
Price/unit		0.293	0.30	0.309	0.317	0.326	0.435	0.505	0.516	0.528	0.541
Deflator $\hat{i}$ = 7%	1.00	1.07	1.14	1.23	1.31	1.40	1.50	1.61	1.72	1.84	1.97
Real price/unit		0.274	0.263	0.251	0.242	0.233	0.290	0.314	0.300	0.287	0.275

$$\left(\sum_{y=1}^{a} MSP_y \times \frac{1}{(1+dr)^y} \right) \div \left(\sum_{y=1}^{a} \frac{1}{(1+dr)^y} \right) = LPP \qquad (2\text{-}25)$$

Where LPP is the levelized production price or levelized cost and MSP is as defined above. This approach to levelized production prices or costs is similar to the calculation techniques shown by Leung and Durning [10], Jeynes [9], and others. Its difference is in the direct application of pro forma statements to levelized cost calculation.

This pricing application of the DCF model is critical to the assessment of new technologies as will be demonstrated in Chapters V-VIII. It is adapted from the revenue requirements approach used for economic analysis by investor owned electric utilities (see, for example, Gulbrand and Leung [6]). Pricing, used as a comparative tool, can be employed to determine:

(1) the potential for new technology introduction, and

(2) the potential gains available from bringing a new technology from pioneer status to maturation.

If a new technology cannot compete with existing processes on a price basis, despite carrying a risk adjusted discount rate, it probably will not be introduced. Similarly the price differential between the production of a product under risk and mature conditions represents the gains available from bringing a new process to maturity.

III. THE USES OF THE DCF MODEL IN NEW TECHNOLOGY ASSESSMENT

The DCF model can be used to make the following assessments of new technologies:

(1) go/no-go decisions,

(2) pricing calculations, and

(3) some comparative assessments.

Analysis of projects involving new technologies, products, or services, is carried out in the manner presented above: the technique for evaluating mature technologies. New technologies differ from mature technologies, however, in that they invariably involve a higher degree of risk. Risk implicitly is incorporated into the analysis of mature technologies through the discount rate. The costs of debt and

of equity are based upon assessments by lenders and equity
investors concerning the risk level for the firm as a whole.
These costs usually reflect previous and proposed investments
in mature technologies. The unusual risks associated with
investment in new technologies are discussed below as they
impact DCF analysis.

A. The Treatment of Risk

DCF flow analyses are common in the evaluation of all new
technologies, from ethanol production to wood derived cattle
food supplements. In many published analyses considerable
attention is paid to evaluating debt/equity ratios and other
nuances in the use of the DCF model. Frequently, however,
there is no explicit or implicit adjustment of the technique
to accommodate the risk associated with new technology.

Numerous techniques for the treatment of risk are debated
in the literature. The use of these techniques, however,
depends upon defining certain classes of risk as follows:

(1) Financial risk, or the degree to which a project is
    leveraged (financed by debt);

(2) Business risk, which includes the potential for
    recession, the ability of the firm's managers to oper-
    ate the company successfully, the potential for regu-
    latory changes (e.g., environmental regulations), and
    market acceptance of a given product (in the case of
    a new product development); and

(3) Technological risk (a form of business risk), or the
    ability of a given new process to operate in the
    manner predicted and specified by the design.

In this text the issue of financial risk is treated by
calculating industrywide discount rates, and by assuming normal debt/equity ratios. Financial risk is not expected to
increase, per se, as new technologies are introduced. The
same cannot be said for business risk and technological risk.

1. Business Risk Analysis. Business risk incorporates a
wide variety of variables as mentioned above. With respect
to these analyses, however, the firm's overall business risk
is changed when a new product is produced and delivered to
the marketplace. Additional business risk is incurred because it is unknown whether a sufficient quantity of the new
product produced can be sold at a price that produces an
acceptable return on the investor's capital contribution.
This condition is known as "market acceptance." Some of the

potential processes and products that fit this category of
new wood processing technologies are: (1) cattle food supplements, (2) human food supplements and substitutes, and (3)
chemicals from wood (e.g., C_1 chemicals). Market acceptance
impacts all such products. Three techniques have been proposed to treat such risk: sensitivity analysis, probability
analysis, and risk adjusted discount rates. Each is discussed below.

Sensitivity analysis involves changing various estimates
or assumptions used in any capital investment analysis, and
determining the impact of such changes on the calculated NPV,
IRR, or revenue requirement [6]. This sensitivity analysis
provides some indication of the effect that assumed values
can have on the decision to accept or reject the project.
Sensitivity analysis does not include the assignment of probabilities to the key variables being examined. Although the
effect of a change in a key variable can be observed, the
probability of that change is not known. Sensitivity analysis is incomplete in that it fails to indicate the likelihood
of various outcomes, and therefore it will be used to a very
limited extent in this text.

Business risk also has been analyzed through the use of
probability analysis [8] [11]. A probability distribution is
developed with the expected value presented as a base case.
Cumulative frequency distributions also are frequently presented. Results can be shown in the form of "there is a 60%
probability that the NPV will be at least $250,000."

One disadvantage of probability analysis is that it requires the setting of probabilities in a subjective manner.
Further these probabilities must be assigned to a large number of variables. A vast computer program is required, therefore, to carry out the numerous calculations (see Hertz, [8])
and a wide variety of outcomes result from the computer runs.
In addition, the frequency distribution outputs from such
programs do not provide the analyst with clear data regarding
go/no-go decisions or straightforward criteria for alternative investment selection. For these reasons, the probability analysis approach is not employed in this text.

Other methods of risk assessment include the certainty
equivalence (CE) as discussed in Schall and Haley [14] and
the risk adjusted discount rate (RADR) methods. The CE
method recognizes that the use of a discount rate combines
the separate variables of time and risk into a single para-
meter (the discount rate). This single variable implies, at
least mathematically, that risk increases over time. While
risk probably does increase over (projected future) time for
most projects, the risk level of some projects can be shown
to be decreasing with time. The CE method therefore has
merit, particularly in the evaluation of projects where risk

is known or suspected to decrease over time. Such is not the case with new technologies, however. For this reason, the CE method is not employed in this text.

The risk adjusted discount rate (RADR) is used in this text to analyze the first capital investments or pioneer plants designed to improve the efficiency of serving an existing market or to penetrate a new market. It replicates market conditions as we know them in that the lenders and equity investors (e.g., venture capital firms) do charge a risk premium based upon the investment uncertainties. The RADR will always be higher than the conventional discount rate. This situation is not unlike firms installing cogeneration before 1978 and the advent of the Public Utilities Regulatory Policies Act (PURPA). The market uncertainties included, among other factors, the potential for being regulated as a public utility. In such cases potential cogenerators in the manufacturing community typically increased their required discount rate by as much as a factor of two over their conventional rate in order to cover the increased marketplace risk [16].

The technique used here, then, is to analyze new technologies, and the capital investments associated with such projects, by using a RADR. Such an approach permits convenient assessment of risks and opportunities associated with bringing a product into a new market. It simulates actual investor behavior. Finally, it provides for both go/no-go type assessments and market analyses by the use of the DCF model in a pricing mode.

2. Technical Risk. Technical risk is a subset of business risk, and is that risk associated with bringing a new or substantially modified process (e.g., organisolv pulping or autocausticizing) to the marketplace. Technical risk may be independent of, or associated with, the market risk associated with bringing a new product (e.g., cattle food) to the marketplace.

There are definite learning curves associated with the development of new technology. These learning curves begin when the technology is at the concept stage existing only in the mind of the scientist. They proceed through the stages of process development unit, pilot plant, pioneer plant, and subsequent commercial plants. Swabb [17] has defined the pioneer plant as the first plant approaching commercial scale, owned by private industry and selling its products in the marketplace. It is the break point between the research and development stage and the commercial deployment stage.

Technical learning curves have associated financial learning curves. The financial learning curves correspond to the degree of confidence that investors can place in a new technology, and its ability to function as predicted by engineers

and technologists. The development of such learning curves
leads to an empirical technique for deriving RADR values used
in assessing technical risk. This technique offers a promising approach to the evaluation of gains potentially available
from new technologies. Consequently it will be used in this
text.

B. The Use of Risk Adjusted Discount Rates

Since greater risk investments must provide the opportunity for greater rewards to the investor, the RADR is used in
place of the mature technology discount rate in all new technology NPV and pricing analyses. The new process investment
must bear a higher annual capital cost (or capital recovery)
factor. Thus, in equations (2-4) through (2-7), RADR will
be substituted for dr. A similar substitution will be made
in equations (2-22) through (2-24).

IV. CONCLUSION

The DCF model offers a highly appropriate methodology for
analyzing new technologies in the Forest Products Industry.
It is useful not only in evaluating go/no-go decisions with
and without unusual risk, but also can be used to accomplish
comparative assessments through levelized pricing. In that
mode, it can be used to assess the market penetration potential of new technologies. Finally, and in the context of new
technology assessment, it can be adjusted to incorporate elements of business and technical risk that must be dealt with
in the evaluation of new processes and products.

REFERENCES

1. Bierman, H., and Smidt, S. 1971. "The Capital Budget-
    ing Decision." 3rd Ed. The Macmillan Co., New York.

2. Blankenship, J. E. 1978. Discounted Cash Flow and
    the Reinvestment Assumption. *Managerial Planning,* May/
    June, 16-38.

3. Copeland, T. E., and Weston, F. J. 1979. "Financial
    Theory and Corporate Policy." Addison-Wesley, Boston,
    Massachusetts.

4. Electric Power Research Institute. 1979. Technology Assessment Guide. EPRI, Palo Alto, California.

5. Granskog, J. E. 1978. Economies of Scale and Trends in the Size of Southern Forest Industries. *In* "Complete Tree Utilization of Southern Pine," Proc. (C. W. McMillin, ed.). Forest Products Research Society, Madison, Wisconsin.

6. Gulbrand, K. A., and Leung, P. 1974. Power System Economics: A Sensitivity Analysis of Annual Fixed Charges. American Society of Mechanical Engineers, New York.

7. Haley, C. W., and Schall, L. D. 1979. "The Theory of Financial Decisions." McGraw-Hill Book Co., New York.

8. Hertz, D. B. 1964. Risk Analysis in Capital Investment. *Harvard Business Review, 42*(1), 117-128.

9. Jeynes, P. H. 1968. "Profitability and Economic Choice." Iowa State University Press, Ames, Iowa.

10. Leung, P., and Durning, R. F. 1977. Power System Economics: On Selection of Engineering Alternatives. American Society of Mechanical Engineers, New York.

11. MicroEconomic Associates. 1978. Effects of Risk on Prices and Quantities of Energy Supplies. EPRI, Palo Alto, California.

12. Neuman, D. G. 1977. "Engineering Economic Analysis," Revised Edition. Engineering Press, San Jose, California.

13. Public Law 97-34. Economic Recovery Tax Act of 1981. August 13, 1981. U.S. Congress.

14. Schall, L. E., and Haley, C. W. 1977. "Introduction to Financial Management." McGraw-Hill Book Co., New York.

15. Seed, A. H. 1982. Inflation Accounting: How It Aids the Financial Executive. *Financial Executive*, L(12): 15-24.

16. Stobaugh, R., and Yergin, D. 1979. "Energy Future: Report of the Energy Project at the Harvard Business School." Random House, New York.

17. Swabb, L. E. 1978. Liquid Fuels from Coal: From R & D to An Industry. *Science 199*(4329):619-622.

18.   Todd, K. R.   1982.   How One Financial Officer Uses
      Inflation-Adjusted Accounting Data.   *Financial Executive*
      L(10):13-19.

19.   U. S. Department of Commerce.   1984.   Statistical
      Abstract.   U. S. Government Printing Office, Washington,
      D.C.

Chapter III

THE DISCOUNT RATE FOR CURRENT FOREST
INDUSTRY INVESTMENTS

I. INTRODUCTION

The economic model for the analysis of capital investments presented in Chapter II relies upon the development of
numerous parameters such as capital costs, revenues, operating and maintenance costs, income tax factors, and discount
rates. Of these variables, revenues are determined by the
market price for the product or products; costs are determined by factors inherent to the technology, and the engineering of that technology; and the tax factors are set by
law. Discount rates, however, are determined independently.
Because discount rates are calculated independently, and because they are perhaps the most critical variables for assessing new technologies, they are treated separately in this
chapter and in Chapter IV.

This chapter provides a careful examination of the discount rate for the Forest Products Industry (FPI) assuming
current conditions and technologies. It includes a calculation of the current discount rate and an examination of historical and technical factors determining that rate. The
discount rate provided in this chapter is appropriate for
evaluating mature technology FPI investments such as the
Leaf River pulp mill depicted in Fig. 1. Further, it is useful as a basis for calculating risk adjusted discount rates,
as presented in Chapter IV.

A. *Calculating the Overall Discount Rate*

Theoretically the discount rate is considered to have
three components [1]: inflation (Inf), premium for early
versus later availability and use of funds (M), and risk (R).
Mathematically it is expressed as follows:

$$DR = Inf + M + R \tag{3-1}$$

Figure 1. Overview of the bleached kraft pulp mill built by Leaf River Forest Products, a company of Great Northern Nekoosa Corporation. This mill demonstrates the type of investment justified by mature technology discount rates. Photo courtesy of Leaf River Forest Products.

The relationship among these variables is treated most frequently as a multiplicative rather than an additive operation (see the Chapter II discussion of nominal and real discount rates). This difference in mathematical treatments, however, is inconsequential (e.g., <5% difference) under most conditions.[1]

Equation (3-1) illustrates the theoretical components of the discount rate. That equation is difficult to use directly in developing a discount rate for any company since M and R are unknown. More empirical means usually are employed to develop a company or industry discount rate, including the weighted average cost of capital (WACC) technique. The WACC equation is shown in equation (3-2) [10]:

$$WACC = k_d \, \theta + (1 - \theta)k_e \qquad (3-2)$$

Where θ is the proportion of capital financed by debt, k_d is the cost of debt, and k_e is the cost of equity.

There are other, more subjective techniques for setting the discount rate or cost of capital such as company surveys of discount rates currently in use. Such techniques have the advantage of incorporating the objectives of the corporation into the value determined, with the objectives being expressed either explicitly or implicitly. Such techniques also provide an accommodation for the influence of heavy competition for capital (e.g., by government borrowing) on discount rates. The WACC technique is preferred here, however, for the following reasons:

(1) it is driven by the market for capital, and therefore reflects the expectations of stockholders and bondholders;

(2) changing conditions in debt and equity markets are reflected rapidly in any values proposed, and

(3) information required to calculate the cost of capital is readily available because the costs and values of the instruments of debt and equity capital are published by such services as Standard and Poors, Moody's Bond Rating Service, Value Line, in the daily newspapers, and in corporate annual reports.

---

[1]*Assume, for the moment, that I = .06, M = .05, and R = .03. Using equation (3-1), the discount rate is .14, or 14%. Using the multiplicative approach, the expression would be: (1.06 × 1.05 × 1.03) - 1 = 1.146 and the consequent discount rate would be 14.6%. The percentage difference is .006/.14 × 100 or 4.3%.*

B. Calculating the Components of the Discount Rate.

Because the WACC approach shown in equation (3-2) is preferred here, the three variables of the equation merit definition: θ, k_d, and k_e. Their definitions can be expressed
mathematically as is shown below.

1. Defining θ. The factor θ, that is the ratio of a
corporation's long term debt to its total long term capital
base, is calculated as follows:

$$\theta = \frac{P_d \times Q_d}{(P_d \times Q_d) + (P_e \times Q_e)} \qquad (3\text{-}3)$$

Where P_d reflects current market prices on long term bonds,
Q_d is the number of bonds outstanding, P_e is the current common stock price, and Q_e is the current number of common stock
shares outstanding (being traded publicly on the New York
Stock Exchange or other markets). For many corporations,
several bond issues may be outstanding at any given time. In
those cases the expression $P_d \times Q_d$ is expanded as follows:

$$P_d \times Q_d = \sum_{i=1}^{n} P_{d_i} \times Q_{d_i} \qquad (3\text{-}4)$$

This expression accumulates the contribution of each bond
issue to the debt portion of the capital structure.

Expression (3-2) relates to companies with no preferred
stock. Numerous corporations have issued preferred stock to
varying degrees. In economic sectors such as the electric
utility industry, preferred stock may comprise a significant
share of the capital structure (e.g., 10 to 15% of the total
capital base). To include preferred stock into the overall
WACC expression, the proportions of debt, preferred stock,
and common stock must be calculated. The cost and structure
of preferred stock is usually somewhere in between that associated with debt and that associated with equity. In manufacturing firms, preferred stock generally is an insignifi-
cant portion of the total capital structure. For these
reasons the capital structure can be divided proportionally
between long term debt and common stock equity without significantly influencing the calculated discount rate, and therefore this two term approach is taken here.

2. Defining the Cost of Debt (k_d). With θ defined, it
is now necessary to determine k_d, the cost of long term debt.
The real cost of debt is the yield to maturity (YTM) on outstanding bonds, since this is approximately the cost of

newly issued debt. The k_d term is calculated as follows:

$$k_d = \text{YTM} \times (1 - \text{TR}) \tag{3-5}$$

Where TR is the incremental income tax rate of the firm
expressed as a decimal. Yield to maturity is calculated as
the current yield of the bond (bond interest payment divided
by the market price of the bond) plus the interest rate that
converts the current bond price into the face value of the
bond over its remaining life. The calculation of YTM is best
shown by an example, assuming a bond with a face yield of
4.5%, a face value of $100, a market price of $85.50, and a
remaining life of 10 years. The current yield is:

$$\frac{\$4.50}{\$85.50} = .0526$$

And the yield from the appreciation of the current market
price to the face value of the bond is:

$$(\$85.5)(1 + i)^{10} = \$100$$

$$i = 0.0156$$

The yield to maturity is the sum of the two yields, or 6.8%.
If the corporation has an incremental tax rate of 35%, then
k_d is as follows:

$$k_d = .0682 \times (1 - .35) = .0443 \quad \text{or } 4.3\%$$

The tax rate deduction is taken because interest expenses are
allowed as a deduction by the IRS in the computation of
income taxes.

   3. Defining the Cost of Equity (k_e). The cost of equity
capital, k_e, can be calculated by several methods including
the Gordon-Shapiro "dividend-growth model" and the "capital
asset pricing model." The dividend-growth model (DGM) is as
follows [4]:

$$k_e = \frac{D}{P} + G \tag{3-6}$$

Where D is the expected annual dividend on the stock over the
next twelve months, P is the current common stock price, and
G is the expected growth rate in dividends of the stock over
an essentially infinite period.
   The Capital Asset Pricing Model formula is stated as
follows:

$$k_e = k_{rf} + \beta_j \ (k_m - k_{rf}) \hspace{3cm} (3\text{-}7)$$

Where k_{rf} is the cost of risk free money (e.g., 90 day treasury bills), k_m is the average return on equity in the market, and β_j is a measure of systematic risk of the firm. The β_j term is defined by Mullins [7] as follows: a stock with a β_j of 1.00 is a stock with an average amount of systematic risk, and it rises and falls at the same percentage as any broad market index (e.g., the Standard and Poor's 500 stock index). This is a simplified form of the Rosenburg and Guy [9] definition of β_j, where the systematic risk measure is calculated as the covariance between the return from dividends plus change in price per share of a given security relative to the market as a whole. The CAPM gives rise to the Security Market Line shown in Fig. 2

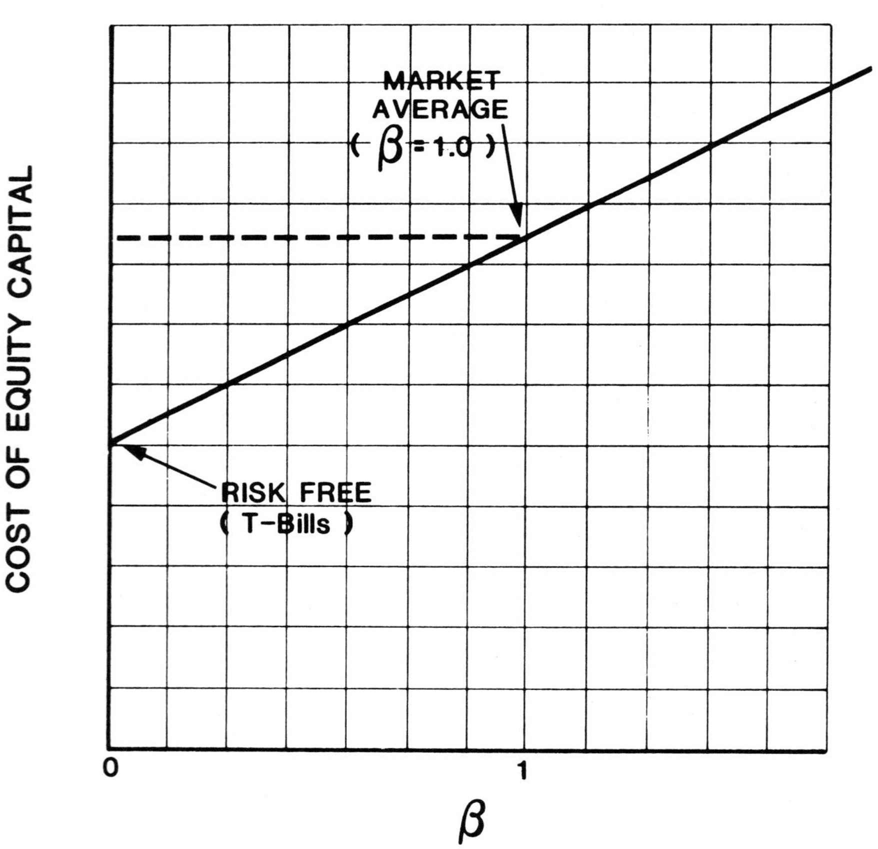

Figure 2. The Security Market Line derived from the Capital Asset Pricing Model. Source: [10].

The DGM and CAPM models yield comparable results under normal market conditions [4], within the limitations of the DGM. Limitations of the DGM include the fact that it assumes perpetual growth of a firm; and these limitations also include the fact that the DGM cannot be used to calculate k_e for companies with unstable dividend payments, no dividend payments, or high growth rates [7]. The DGM also tends to break down during periods of high inflation. These limitations make the Gordon-Shapiro model less applicable than the CAPM [7]. The CAPM has additional advantages including the explicit treatment of risk and a close relationship to the theoretical discount rate as expressed in equation (3-1). The limitations of the DGM and the advantages of the CAPM make the latter more appropriate here.[1]

C. Conclusions Concerning Discount Rate Calculation

As can be seen from the models and equations presented above, each firm has its own cost of capital or discount rate. That corporate discount rate reflects the assessment of corporate risk by the holders of capital instruments (bonds and stocks). Bond prices decline (hence yields to maturity increase) as bonds decrease in rating from AAA to C on the Standard and Poor's rating system, or decline comparably on the Moody's scale. The cost of equity increases linearly as a function of β_j, or the risk assumed in achieving expected dividend payments. This risk-capital cost relationship extends beyond publicly traded stocks and into nontraded capital markets as well (e.g., privately held corporations and venture capital). Typically discount rates of firms within an industry are sufficiently similar that an industry wide discount rate can be approximated. Such a cost of capital is presented below.

II. THE DISCOUNT RATE OF THE FOREST PRODUCTS INDUSTRY

With that brief summary of the theory of discount rate setting, it is now possible to calculate a discount rate of the FPI. Such a calculation provides a basis for (1) an analysis of historical trends associated with the FPI discount rate, (2) an assessment of technical influences on the FPI discount rate, and (3) a basis for setting discount

---

[1]*In one case both DGM and CAPM values will be shown for comparison purposes.*

rates appropriate for new and novel wood processing technologies.
   In order to determine the cost of capital in the FPI, a
sample of publicly traded firms was chosen. This sample was
then evaluated in terms of the three key variables: θ, k_d,
and k_e. The list of firms in the sample is as follows:
Boise Cascade, Champion International, Consolidated Papers,
Crown Zellerbach, Fort Howard Paper, Georgia Pacific, Great
Northern Nekoosa, International Paper, Kimberly Clark, Mead
Corp., St. Regis Corp., Scott Paper, Union Camp, Westvaco,
and Weyerhaeuser. These firms share the following characteristics:

   (1) they are traditional forest products firms, rather
       than broad based companies with significant forest
       industry activities (e.g., Koppers, Mobil);

   (2) over 90% of their total capital structure is publicly
       traded, including significant proportions of their
       debt capital; and

   (3) taken as a group they produce the full range of
       forest products, in varying quantities.

The 1983 discount rate for these firms is shown in
Table I. General values concerning the Forest Industry discount rate can be derived from Table I, and these general
values are shown in Table II.
   The overall nominal discount rate for the Forest Products
Industry is about 15% (14.5%), as is shown in the tables
above. This cost of capital is somewhat higher than the discount rate associated with the marketplace as a whole, and it
is based apparently upon a perception by investors that risk
in the FPI carries somewhat more risk than the average for
all industries. Given an inflation rate currently estimated
at about 5% (4.5%), the real discount rate of the FPI is
about 10% (9.5%).
   Of final interest in calculating the current discount
rate of the FPI is the influence of model selection on the
estimate of the cost of equity (k_e). This comparison of the
DGM and CAPM models is shown in Table III, based upon data
obtained from the Value Line Investment Survey [23]. The
cost of equity for forest industry firms can be approximated
by DGM because such firms are large and are slow growing.
Further such firms consistently pay dividends that do not
change erratically. Consequently the FPI does not violate
the conditions necessary to use the Gordon-Shapiro model as
described by Mullins [7]. Note that, on average, the two
models produce k_e values that are within 8% of each other
((.175-.162)/.162 = .08), with the dividend-growth model

TABLE I. The 1983 Discount Rate for Fifteen Forest Products Industry Companies
(Decimal basis)

Company	θ	Yield to maturity	Tax rate	Cost of debt	β_j	Cost of[a] equity	Discount rate
Boise Cascade	.37	.112	.35	.073	1.15	.182	.142
Champion International	.39	.112	.32	.076	1.20	.187	.144
Consolidated Papers	.05	N/A	.45	N/A	0.85	.156	.156
Crown Zellerbach	.40	.125	.35	.081	1.20	.187	.145
Fort Howard Paper	.08	N/A	.45	N/A	0.90	.160	.160
Georgia Pacific	.45	.115	.36	.074	1.25	.191	.138
Great Northern Nekoosa	.31	.105	.35	.068	1.05	.173	.141
International Paper	.24	.111	.30	.078	1.10	.178	.154
Kimberly Clark	.20	.107	.35	.069	0.85	.156	.138
Mead Corp.	.53	.123	.35	.080	1.15	.182	.128
St. Regis	.38	.119	.30	.083	1.10	.178	.142
Scott Paper	.26	.119	.35	.077	1.00	.169	.145
Union Camp	.29	.126	.35	.082	0.95	.165	.141
Westvaco	.27	.123	.37	.077	1.10	.178	.151
Weyerhaeuser	.35	.110	.32	.075	1.20	.187	.148

[a]This assumes a risk premium (k_M-k_{RF}) of 8.9% and a k_{RF} = .080. Sources: Mullins [7], Value Line [23], Standard and Poors [11].

TABLE II.   General Discount Rate Values for the Sample Forest Products Industry Firms

		Value	
Parameter	*Mean*	*Standard deviation*	*Range*
Proportion of debt (θ)	*30.5%*	*12.9%*	*48.0%*
Yield to maturity[a]	*11.6%*	*0.7%*	*2.1%*
Tax rate	*35.5%*	*4.4%*	*15.0%*
Cost of debt[a]	*7.6%*	*0.5%*	*1.5%*
Risk factor (β_j)	*1.07*	*0.13*	*0.40*
Cost of equity	*17.5%*	*1.2%*	*3.5%*
Cost of capital	*14.5%*	*0.8%*	*3.2%*

[a]*Does not include Consolidated Papers and Fort Howard Paper.   Source: Table I.*

producing only a slightly lower value. It is recognized that the results from DGM are more volatile than those associated with CAPM, as is shown in Table III. The comparability of k_e values as a function of model selection was evaluated through the use of the student-t test. The application of the student-t test results in a t statistic of -1.59. Since the critical t value at 14 degrees of freedom is +2.624 (99% confidence interval), it can be seen that model selection does not significantly influence the calculated k_e value.

III. HISTORICAL TRENDS IN THE DISCOUNT RATE OF THE FOREST PRODUCTS INDUSTRY

To understand the interplay between new technologies and the FPI discount rate, it is essential to go beyond the simple presentation of the current industry cost of capital and examine trends in the discount rate over time. Further it is essential to evaluate historical and technical forces influencing the cost of capital in the FPI in recent years. Historical trends are examined in this section while technical forces are reviewed in Section IV. In reviewing historical trends, the cost of debt and the cost of equity are treated separately.

TABLE III. A Comparison of Models Used to Calculate the Cost of Equity in the FPI (Value Line [23])

Company	D/P	G	$k_{e_{(D/P)}}$	β_j	$k_{e_{(CAPM)}}$
Boise Cascade	.054	.140	.194	1.15	.182
Champion International	.042	.160	.202	1.20	.187
Consolidated Papers	.063	.080	.143	0.85	.156
Crown Zellerbach	.075	.175	.250	1.20	.187
Fort Howard Paper	.024	.110	.133	0.90	.160
Georgia Pacific	.045	.135	.180	1.25	.191
Great Northern Nekoosa	.058	.075	.133	1.05	.173
International Paper	.073	.080	.153	1.10	.178
Kimberly Clark	.059	.055	.114	0.85	.156
Mead Corp.	.093	.125	.218	1.15	.182
St. Regis	.070	.065	.135	1.10	.178
Scott Paper	.054	.075	.129	1.00	.169
Union Camp	.050	.065	.115	0.95	.165
Westvaco	.047	.100	.147	1.10	.178
Weyerhaeuser	.037	.145	.182	1.20	.187
Average	.056	.106	.162	1.07	.175
Standard Deviation	–	–	.040	–	.012

A. *The Historical Cost of Debt in the Forest Products*
    Industry

In order to evaluate the cost of debt over time in the
FPI, a sample of nine firms with publicly traded debt was
selected. Data from these firms are presented for the years
1972, 1976, and 1980-1983 in Table IV, based upon Standard
and Poor's data [11]. It is significant to note that, over
time, the real cost of debt has increased significantly in
the FPI. Because the cost of debt has increased in real
terms over time, it is necessary to evaluate the contribution
of risk to that price rise. Such an evaluation is shown in
Table V. The current risk contribution to the cost of debt
is 4.4% (real).

B. *Historical Costs of Equity in the Forest Products*
    Industry

Because of the structure of the CAPM, the critical variable to evaluate with respect to time is the β_j term. Other
variables, on a real basis, can be netted out as a constant.
The β_j term for the nine firms analyzed has been averaged for
every year since 1972, based upon data from Value Line
Investment Survey [12-23]. A plot of these data is shown in
Fig. 3. Note that there was a tremendous increase in perceived risk between the years 1972 and 1973. In that time
frame the risk associated with the FPI changed from being
less than the risk associated with the market as a whole to
being greater than the risk associated with a total market
portfolio.
Since 1973 there has been a slight upward drift in the
FPI term [13-23]. This drift conforms generally to the
equation:

$$\beta_j = 0.0024A - 3.672 \tag{3-8}$$

Where A is the calendar year (e.g., 1983). The trend line is
also shown in Fig. 3. This indicates that there has been
somewhat of an increase in the perceived investor risk in the
FPI during the past decade. This perception has made a small
contribution of about 0.3% to the real cost of equity.

C. *Historical Trends in the Cost of Capital for the*
    Forest Products Industry

The rising costs of debt and equity have, necessarily,
led to a rise in the cost of capital for the FPI. This rise

TABLE IV.   Historical Data on Bond Yields to Maturity in the Forest Products Industry
(% Basis)

Firm	1972	1976	1979	1980	1981	1982	1983
Crown Zellerbach	7.96	9.15	9.86	12.96	13.39	16.99	12.80
Georgia Pacific	9.45	13.01	10.15	14.50	17.76	17.82	11.65
International Paper	8.20	8.75	9.74	12.88	12.85	15.08	11.80
Kimberly Clark	7.07	8.24	7.36	9.84	11.40	10.81	11.70
Mead Corp.	8.21	10.43	10.44	14.46	14.48	16.44	14.86
Scott Paper	7.98	8.95	10.22	14.27	14.56	16.09	12.52
Union Camp	7.54	9.21	8.90	11.45	13.64	16.91	12.68
Westvaco	8.14	14.31	9.44	13.51	15.27	17.03	13.62
Weyerhaeuser	7.86	8.52	8.86	13.01	13.81	15.27	12.12
Mean	8.05	10.06	9.66	12.99	14.13	15.83	12.64
Inflation	4.2	5.2	8.6	9.3	9.4	6.7	4.5
Real mean yield to maturity	3.7	4.6	1.0	3.4	4.3	8.6	7.8

Source: Standard and Poors [11].

TABLE V.   The Risk Free and FPI Costs of Borrowing Over Time (Values in percent)

| Year | Nominal costs | | | Real costs | | Estimated risk premium[b] |
	Govt[a]	FPI	Inflation	Govt	FPI	
1972	6.3	8.1	4.2	2.0	3.7	1.7
1976	6.8	10.1	5.2	1.5	4.6	3.0
1979	8.7	9.7	8.6	0.1	1.0	0.9
1980	10.8	13.0	9.3	1.4	3.4	2.0
1981	12.9	14.1	9.4	3.2	4.3	1.1
1982	12.7	15.8	6.7	5.6	8.6	2.8
1983	8.0	12.6	4.5	3.8	7.8	4.4

[a] *Long term bonds, representative April values.*

[b] *Estimated as Real Cost, FPI ÷ Real Cost, Govt.*

has occurred on both a nominal and a real basis. Much of this rise has been consistent with the market as a whole, for capital is now more expensive than it has been at least since World War II. However some of the rise in the FPI can be attributed to increased perceptions of risk by investors.

The rise in the cost of capital in the FPI has been dramatic within the past decade. As Fig. 4 illustrates, the real cost of capital has risen from about 4% to 10%. This rise has been measured at 3-year intervals to remove the undue influences of the severe recession in 1981–1982. The implications of Fig. 4 are severe. As the FPI approaches plant investment for modernization and increased market penetration, it must do so in the face of increasing capital cost. Such capital costs will make new investments increasingly difficult to justify. Such capital costs also will make products from new investments increasingly expensive to produce.

Because the cost of capital in the FPI has been rising in recent years, it is essential to investigate the technical influences on that WACC. It is instructive to compare the discount rate of the FPI to discount rates associated with related industries within the context of a technical analysis. Such evaluations are presented below.

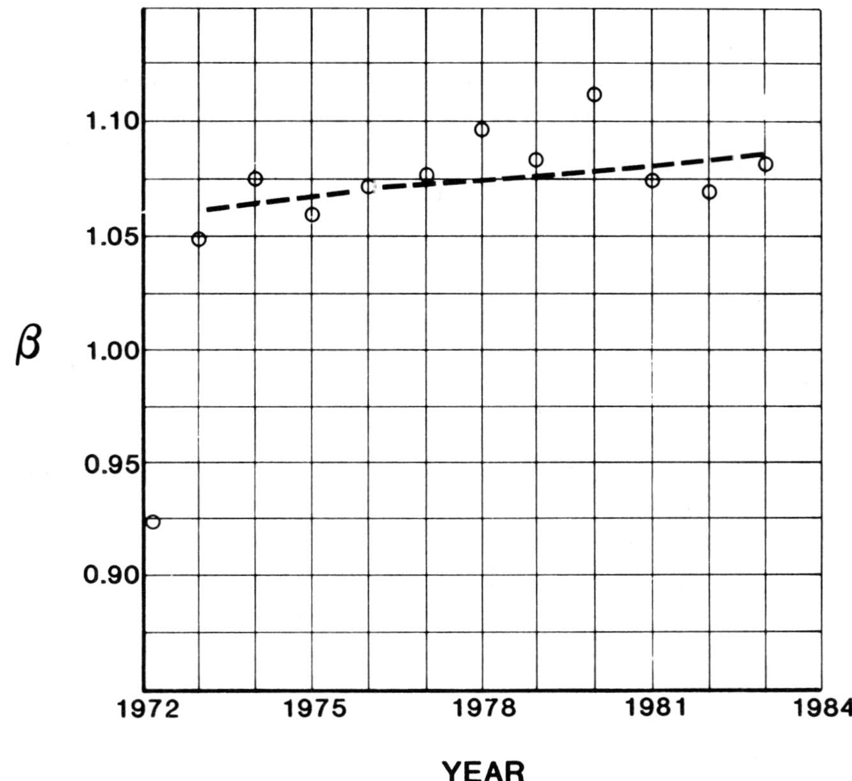

FIGURE 3.   Historical trends in the Beta value of the
Forest Products Industry.   Sources: [11-23].

IV.   TECHNICAL INFLUENCES ON THE DISCOUNT RATE OF THE
      FOREST PRODUCTS INDUSTRY

From a technical perspective there are two issues merit-
ing analysis: (1) what is the influence of debt, or leverage, on the FPI discount rate, and (2) what is the influence
of product mix on the discount rate. Other factors influencing the discount rate may include the degree of timber
self-sufficiency, the size of the corporation as measured
either by assets or sales, and similar factors. The issues
identified above, however, are of most interest here.

*A. The Influence of Debt on the Discount Rate of the
     Forest Products Industry*

The Modigliani-Miller (M-M) theorem states that, if
taxes are ignored, a corporation is indifferent to the method
of financing [5]. This theorem argues that, while the cost
of debt is less than the cost of equity, borrowing increases
the financial risk associated with the equity investment. The
M-M theorem concludes that, in a taxless situation, the WACC
will be the same regardless of the value calculated for θ [5].
Given the influence of tax laws, and the tax deductibility of
interest, debt financing is preferred to some level. Such
debt financing permits the corporation to capture the benefits of lower after-tax costs of debt (relative to equity)
and hopefully the realization of a lower WACC than if debt

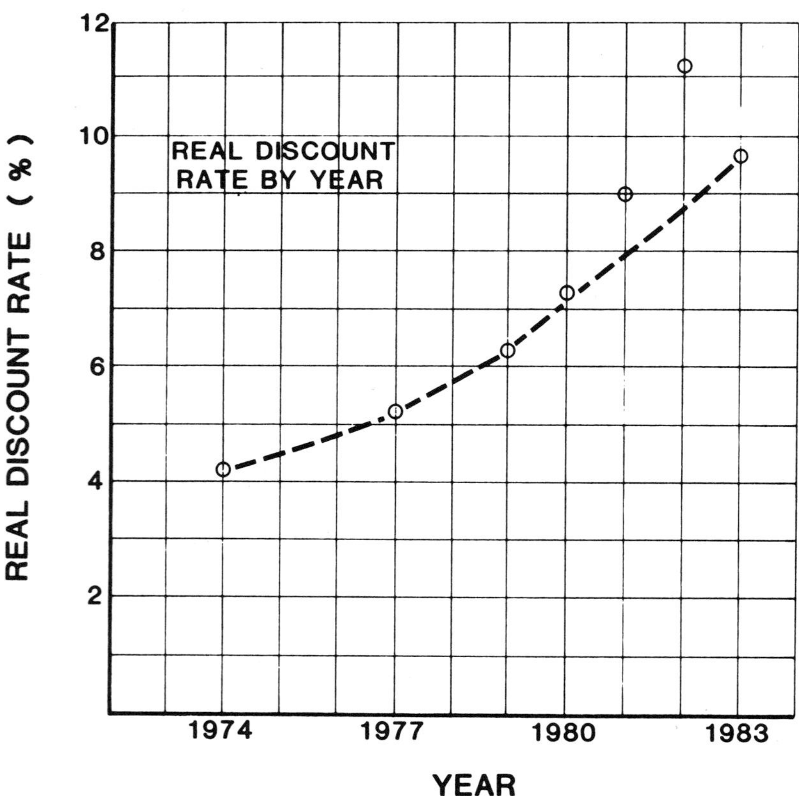

FIGURE 4. *Historical trends in the cost of capital of
the Forest Products Industry. Sources: [11-23].*

had not been employed. It is necessary to evaluate that percentage for the FPI.

It can be seen from Table I that the cost of debt capital as reported by Standard and Poors [11] is significantly less than the cost of equity capital in the FPI. The average cost of long term debt is 7.6% while the average cost of equity is 17.5%. Given this difference it is useful to examine the influence of debt further.

In order to examine the influence of debt, the cost of capital has been calculated assuming no debt financing in the corporation. Two additional formulae have been used to perform this analysis. These formulae, explained in detail by Copeland and Weston [3], disaggregate business and financial risk, and calculate the unleveraged β for the firm.

The CAPM disaggregating business and financial risk becomes:

$$k_e = k_{rf} + \beta_u (k_m - k_{rf}) + \beta_u (k_m + k_{rf}) \left(\frac{\theta(1 - TR)}{1 - \theta} \right)$$

$$(3-9)$$

Where β_u is the beta term for the unleveraged firm (it is equivalent to the beta term for measuring business risk).

The calculation of β_u then becomes:

$$\beta_u = \frac{\beta_j}{1 + \frac{\theta(1 - TR)}{1 - \theta}}$$

$$(3-10)$$

Where β_j is the risk factor for the firm, including debt financing, as presented previously.

The β_u has been calculated for the fifteen FPI firms shown in Table I. These values are shown in Table VI, along with the term and the contribution of business risk to total risk. From these data it can be seen that the average unleveraged beta (β_u) for the FPI is 0.80, and the contribution of business risk to total risk is 75%.

The cost of unleveraged equity capital has been calculated for each firm listed in Table I, based upon the β_u terms shown in Table VI. These calculated values are shown in Table VII, along with the WACC values from Table I. From Table VI it is apparent that debt is now accomplishing only a very modest decrease in the cost of capital experienced by the FPI. This modest influence can be attributed, perhaps, to the following factors:

*TABLE VI.   β Terms for Fifteen FPI Firms, and the
Contribution of Business Risk to Total Risk*

Firm	β_u	β_j	*Percent risk attributed to business*[a]
Boise Cascade	*.83*	*1.15*	72
Champion International	*.84*	*1.20*	70
Consolidated Paper	*.83*	*.85*	98
Crown Zellerbach	*.84*	*1.20*	70
Fort Howard Paper	*.86*	*.96*	96
Georgia Pacific	*.82*	*1.25*	66
Great Northern Nekoosa	*.81*	*1.05*	77
International Paper	*.90*	*1.10*	82
Kimberly Clark	*.73*	*.85*	86
Mead Corp.	*.66*	*1.15*	57
St. Regis	*.77*	*1.10*	70
Scott Paper	*.81*	*1.00*	81
Union Camp	*.75*	*.95*	79
Westvaco	*.73*	*1.10*	66
Weyerhaeuser	*.88*	*1.20*	73
Average (Mean)	*.80*	*1.07*	75

[a] $\beta_u / \beta_j \times 100$.   *Source: Value Line [23].*

(1) The FPI may be near the limit of borrowing as determined by the capital markets (see Table V for possible substantiation); and

(2) The tax impact on the FPI is lower than that of
industry as a whole, due to the capital gains treatment on timber and the ability to attribute some cash
flows to timber harvesting rather than wood processing (typically the FPI has a tax rate of 35%, compared to the marginal Federal corporate income tax
rate of 46%).

*TABLE VII. The Unleveraged and Total Cost of
Capital for Fifteen Forest Industry Firms*

Company	Unleveraged cost of equity capital (%)	WACC (%)	Difference (%)
Boise Cascade	15.4	14.2	1.2
Champion International	15.5	14.4	1.1
Consolidated Paper	15.4	15.6	(0.2)
Crown Zellerbach	15.5	14.5	1.0
Fort Howard Paper	15.7	16.0	(0.3)
Georgia Pacific	15.3	13.8	1.5
Great Northern Nekoosa	15.2	14.1	1.1
International Paper	16.0	15.4	0.6
Kimberly Clark	14.5	13.8	0.7
Mead Corp.	13.9	12.8	1.1
St. Regis	14.9	14.2	0.7
Scott Paper	15.2	14.5	0.7
Union Camp	14.7	14.1	0.6
Westvaco	14.5	15.1	(0.6)
Weyerhaeuser	15.8	14.8	1.0
Average (Mean)	15.2	14.5	0.7

The consequence of the data presented in Table VI, however, is that subsequent analyses can be made using either the unleveraged cost of equity capital, or the WACC, with no material impact on the results. Subsequent analyses are indifferent because both values round to a nominal discount rate of 15% or a real discount rate of 10% (assuming a future inflation rate of about 5%).

B. *The Influence of Product Slate on the Discount Rate*
    of the Forest Products Industry

The issue of product slate, as it influences the discount
rate, is far more complex than the issue of financial structure. Firms in the FPI produce a wide variety of solid wood,
pulp, and nonwood products. Whether these varying product
slates measurably influence the discount rate is open to
question. A portion of the question is whether or not these
different products should be treated as a product portfolio
(see Biggadike [2] for a discussion of product portfolios).
Alternatively all forest products may function as if they were
a single product for purposes of the investment community.
Such an alternative is supported at least partially by the
presentation of Mullins [7].

Product slates, β_j terms, and WACC calculations for the
fifteen firm sample are shown in Table VIII. Regression
analysis applied to these values provides two significant
equations:

$$\beta_j = 1.00 + .0043\%SWP \tag{3-11}$$

and,

$$\beta_j = 1.38 - .0045\%PP \tag{3-12}$$

Where %SWP is the percentage of total business in solid wood
products and %PP is the percentage to total business in pulp
and paper. The r values (coefficients of correlation) for
these regression equations are 0.67 and -0.79 respectively.
These are not particularly strong, but are statistically
significant.

Equations (3-11) and (3-12) lead to a simple binomial
expression for calculating β_j as a function of product slate:

$$\beta_j = 0.95P + 1.40S \tag{3-13}$$

Where P is the fraction of business in pulp and paper, and S
is the fraction of business in solid wood products.

The influence of expression (3-13) on corporate costs of
capital is obvious. It may be a partial explanation of the
decision by Scott Paper to concentrate on pulp based businesses, and to leave the solid wood products industry. At
the same time, however, expression (3-13) appears to be a
summary of far more complex phenomena. FPI product slates
may not exert overwhelming influence on the discount rate.
When expressions similar to (3-11) and (3-12) were attempted
on the basis of β_u (the unleveraged beta) terms, no significant correlation was obtained. As a consequence, for

TABLE VIII. Product Slate in the FPI and Weighted Average Cost of Capital

| Company | Percent of product by type[a] | | | β_j | WACC (%) |
	Non-FPI	Pulp and paper	Solid wood product		
Boise Cascade	30	53	17	1.15	14.1
Champion International	0	45	55	1.20	14.4
Consolidated Papers	21	79	0	0.85	15.6
Crown Zellerbach	29	55	16	1.20	14.5
Fort Howard Paper	2	98	0	0.90	16.0
Georgia Pacific	14	22	64	1.25	13.9
Great Northern Nekoosa	4	91	5	1.05	14.6
International Paper	9	78	13	1.10	15.5
Kimberly Clark	1	92	7	0.85	14.0
Mead Corp.	56	40	4	1.15	13.7
St. Regis	18	75	7	1.10	14.5
Scott Paper	8	90	2	1.00	14.6
Union Camp	10	79	11	0.95	14.1
Westvaco	10	89	1	1.10	15.3
Weyerhaeuser	3	52	45	1.20	15.0

[a]*Sources: O'Laughlin and Ellefson [8]; Table I.*

the purposes of these analyses, nominal rates of 15% and real discount rates of 10% are used for all FPI products from mature technologies. Such a treatment is consistent with the analysis of Mullins [7], who estimated a single β_j value (1.2) for all forest industry and paper investments.

V. THE DISCOUNT RATE OF RELATED INDUSTRIES

Chapter I identified two key expansion areas potentially available to the FPI in addition to its current product lines: (1) fuels and energy production for outside markets, and

(2) the production of chemicals and foodstuffs. Discount
rates for these industries are shown below.

It is important to examine these related industry discount
rates, particularly since these are the potential growth
areas for the FPI. The importance of such an examination
stems from the potential influence of such hurdle rates on
the FPI discount rate. Harrington [6] has shown that each
type of business has its own discount rate, and that a corporate discount rate is really a composite of all of the
corporation's divisional discount rates. Harrington [6] used
industry related β_j terms to estimate the cost of capital for
various segments of a diversified firm, Alaska Interstate.
In this text, however, the analysis of related businesses
such as energy and chemicals is useful in determining an FPI
industry-wide discount rate trend.

Discount rates have been calculated for energy and chemicals firms as of 1983, and these calculations are presented
in Tables IX (energy) and X (chemicals). It has been proposed that utility discount rates could be used for cogeneration projects, however such rates are not applicable due to
the differences in governmental regulation applied to the
manufacturing and utility economic sectors, including the
degree of competition experienced and the degree of market
diversity allowed.

The results contained in Tables IX and X are summarized
in Table XI. From Table XI it can be seen that the production of fuels can be expected to raise the FPI discount rate
only slightly, if at all. The production of chemicals may
result in a slight decrease in the WACC of forest industry
firms. The changes projected for the FPI discount rate as
calculated and displayed in Table XI are within the error of
estimation for energy production, and <1% lower for chemicals
manufacture. The impact on the cost of capital of expanded
production of electricity as a product for sale cannot be
measured in this way (e.g., by using investor owned electric
utilities as a means for calculating β_j proxy values) due to
structural and institutional differences between the FPI and
the regulated utility industries. If proxy β_j values are to
be calculated, the most appropriate industry for use as a
proxy appears to be petroleum refining due to its heavy current emphasis on cogeneration.

TABLE IX.   The Cost of Capital for Selected Integrated Energy Firms in 1983

Firm	β_j	θ	YTM	1-TR	k_D	$1-\theta$	k_e	WACC
Amerada Hess	1.35	.43	.112	.39	.044	.57	.200	13.3
ARCO	1.20	.36	.124	.56	.069	.64	.187	14.5
Charter Co.	1.45	.45	.146	.45	.066	.55	.209	15.9
Exxon	0.85	.15	.108	.50	.054	.85	.156	14.0
Gulf Oil	1.10	.17	.112	.40	.045	.87	.178	15.5
Mobil Corp.	1.10	.24	.115	.50	.058	.76	.178	14.9
Pennzoil	1.25	.50	.126	.65	.082	.50	.191	13.7
Phillips Petroleum	1.15	.25	.121	.35	.042	.75	.182	14.7
Shell Oil	1.20	.34	.129	.58	.075	.66	.187	14.9
Standard Oil/Calif.	1.10	.14	.113	.60	.068	.86	.178	16.2
Standard Oil/Ind.	1.20	.26	.113	.55	.062	.74	.187	15.4
Standard Oil/Ohio	1.25	.36	.119	.56	.067	.64	.191	14.6
Sun Oil	1.15	.30	.125	.57	.071	.70	.182	14.9
Texaco	0.95	.15	.114	.53	.060	.85	.164	14.9
Average	1.16							14.8

Sources: Value Line [25]; Standard and Poors [11].

TABLE X.   *The Cost of Capital for Selected Basic Chemical Manufacturers in 1983*

Firm	β_j	θ	YTM	1-TR	k_D	$1-\theta$	k_e	WACC
Celanese	0.90	.43	.124	.64	.079	.57	.160	.125
Dow Chemical	1.25	.41	.121	.62	.075	.59	.191	.144
DuPont	1.15	.36	.118	.40	.047	.64	.182	.134
Essex Chemical	0.90	.61	.134	.58	.078	.39	.160	.110
Hercules	1.15	.29	.133	.65	.086	.71	.182	.155
Monsanto	1.00	.22	.119	.61	.073	.78	.169	.148
Rohm and Haas	0.95	.23	.135	.62	.084	.77	.165	.146
Stauffer	1.00	.28	.118	.64	.076	.72	.169	.143
Union Carbide	1.05	.32	.124	.66	.082	.68	.173	.144
Average	1.04							.139

Sources: *Value Line* [24]; *Standard and Poors* [11].

*TABLE XI.   Risk (β_j) Values and Discount Rates
by Industry*

Industry	β_j	Discount rate
Forest products	1.07	14.5%
Energy	1.16	14.8%
Chemicals	1.04	13.9%

VI. CONCLUSION

The data presented above can be summarized in the following salient points:

(1) The current discount rate of the FPI, as calculated by the CAPM or the DGM, is about 15% on a nominal basis or 10% on a real (inflation free) basis.

(2) The real discount rate of the FPI has more than doubled in the past decade, partially as a function of overall market conditions and partially as a function of increased risk perceived by investors.

(3) The increase in the real discount rate is making FPI investments in modernization and diversification more costly and difficult.

(4) The use of debt financing now is having only a modest, if any, influence on the FPI discount rate.

(5) The product slate distributions of the FPI may or may not be direct determinants of the discount rate, but probably reflect more complex phenomena.

(6) Readily alternative businesses available to the FPI will have only a modest impact on the discount rate.

These conclusions apply to investments in proven technologies. In such cases the FPI will experience a nominal discount rate of about 15%, 1% above the market as a whole. The discount rates appropriate for advanced technologies may be substantially different, as will be seen in Chapter IV.

REFERENCES

1. Alchian, A., and Allen, W. R. 1977. "Exchange and Production: Competition, Coordination, and Control." 2nd Ed. Wadsworth Publishing Co., Belmont, California

2. Biggadike, R. 1979. The Risky Business of Diversification. *Harvard Business Review 57*(3):103-111.

3. Copeland, T. E., and Weston, J. F. 1979. "Financial Theory and Corporate Policy." Addison-Wesley, Reading, Massachusetts.

4. Gordon, M. J., and Halpern, P. 1979. Cost of Capital for a Division of a Firm. *Journal of Finance 29*(4): 1153-1163.

5. Haley, C. W., and Schall, L. D. 1979. "The Theory of Financial Decisions." McGraw-Hill, New York.

6. Harrington, D. R. 1983. Stock Prices, Beta, and Strategic Planning. *Harvard Business Review 61*(3): 157-163.

7. Mullins, D. W. 1982. Does the Capital Asset Pricing Model Work? *Harvard Business Review 60*(1):105-114.

8. O'Laughlin, J., and Ellefson, P. V. 1982. New Diversified Entrants Among U.S. Wood-Based Companies: A Study of Economic Structure and Corporate Strategy. Agricultural Experiment Station, University of Minnesota, St. Paul, Minnesota (Station Bulletin 541 - 1982).

9. Rosenberg, B., and Guy, J. 1976. Prediction of Beta from Investment Fundamentals. *Financial Analysis Journal,* May/June, 60-70.

10. Schall, L. D., and Haley, C. W. 1977. "Introduction to Financial Management." McGraw-Hill, New York.

11. Standard and Poors. 1983. Standard and Poors Bond Guide, May, New York (and comparable volumes for prior years).

12. Value Line. 1972. Paper and Forest Products Industry. Value Line Investment Survey, May 19, 761-824.

13. Value Line. 1973. Paper and Forest Products Industry. Value Line Investment Survey, May 18, 861-936.

14. Value Line. 1974. Paper and Forest Products Industry. Value Line Investment Survey, May 17, 862-935.

15. Value Line. 1975. Paper and Forest Products Industry. Value Line Investment Survey, May 16, 861-945.

16. Value Line. 1976. Paper and Forest Products Industry.
    Value Line Investment Survey, May 14, 927-952.

17. Value Line. 1977. Paper and Forest Products Industry.
    Value Line Investment Survey, May 13, 931-957.

18. Value Line. 1978. Paper and Forest Products Industry.
    Value Line Investment Survey, May 12, 931-959.

19. Value Line. 1979. Paper and Forest Products Industry.
    Value Line Investment Survey, May 11, 930-957.

20. Value Line. 1980. Paper and Forest Products Industry.
    Value Line Investment Survey, May 9, 924-951.

21. Value Line. 1981. Paper and Forest Products Industry.
    Value Line Investment Survey, May 8, 926-954.

22. Value Line. 1982. Paper and Forest Products Industry.
    Value Line Investment Survey, Feb. 5, 925-954.

23. Value Line. 1983. Paper and Forest Products Industry.
    Value Line Investment Survey, May 6, 926-953.

24. Value Line. 1983. Chemical (Basic) Industry. Value
    Line Investment Survey, May 20, 1237-1249.

25. Value Line. 1983. Petroleum (Integrated) Industry.
    Value Line Investment Survey, July 15, 401-439.

Chapter IV

DISCOUNT RATES FOR NEW TECHNOLOGIES
IN THE FOREST PRODUCTS INDUSTRY

I. INTRODUCTION

Chapter III provided an assessment of the discount rate
for the Forest Products Industry (FPI) under current market
and technology conditions, and discussed a disaggregation of
business risk and financial risk. Chapter III did not
explicitly isolate the components of business risk, however.
Those business risk components include: (1) company management, (2) economic conditions, (3) regulatory conditions, and
(4) technology expectations. The first three elements of business risk are outside the scope of this text. The influence
of technology on business risk and consequently the discount
rate can be substantial, however, and it is the focus of this
chapter.

This measuring of the explicit impact of technology on
the discount rate is necessary for the construction of a risk
adjusted discount rate (RADR) curve. While this approach has
been criticized by some authors (e.g., Robichek and Myers
[29], Bisio and Gastwirt [3]), there is sufficient support
for it in marketplace behavior and theoretical treatises
(e.g., Lindsay and Sometz [22]) that it is used here. Measuring the impact of technologies on the discount rate is
critical to understanding the fate of projects moving
through the process of innovation defined by Cox [7].

The tie between technological knowledge and investment
risk is best illustrated by an example from the kraft pulping
industry. When the Tomlinson furnace for simultaneous energy
and pulping chemicals was invented, it was first installed in
Cornwall, Ontario, in 1926. In that installation three
liquor recovery boilers were built so that .one unit would be
operating at any given time. The Tomlinson recovery boiler
matured during the 1930s, and today the availability of a
single unit typically exceeds 95% [35, 40].

A.   The Basic Principle of Technology Risk Adjustment

The Capital Asset Pricing Model (CAPM) as discussed in
Chapter III provides for risk adjustment based upon a calculation of the beta (β) term. Brealy and Myers [4] maintain
that this β is project-specific, not company-specific and
that a project either is valid or not valid, independent of
the firm making the investment. Their argument appears as a
Project Market Line (PML) as shown in Fig. 1. This PML is
analogous to the Security Market Line of Chapter III. A
given technology may appear anywhere along the PML, depending
upon its degree of risk. Typically it would start in a position at the extreme right of the project market line and move
downward and to the left as the technology matures. That
movement results from progress along two parallel curves:
the technology learning curve and the financial learning

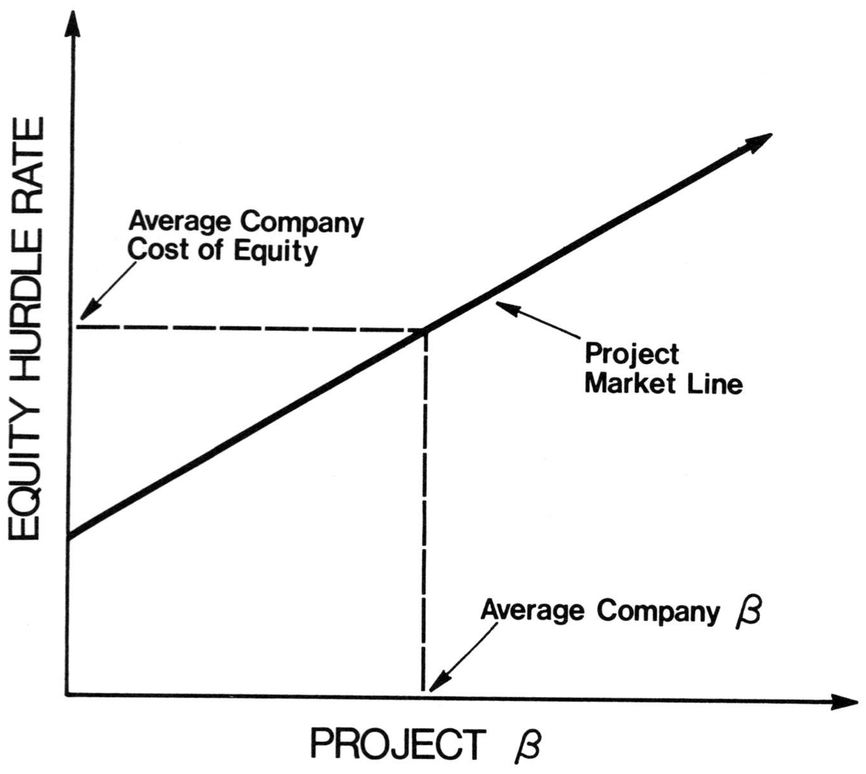

FIGURE 1.   The Project Market Line as adapted from
Brealy and Myers [4].

curve associated with that technology. This PML results from
the assumption that risks associated with any two or more
projects are independent of each other, and the consequent
conclusion that portfolio analysis is inappropriate for project assessment [24].

    1. The Technology Learning Curve. In a paper on tech-
nology commercialization, Swabb [34] identified three critical stages of maturation:

   (1) research and development, leading from bench scale
       process development unit to the large scale pilot
       plant;

   (2) design and construction of the pioneer plant, defined
       as the first commercial scale stand-alone facility,
       intended to be operated as a commercial venture over
       a conventional plant life; and

   (3) design and construction of subsequent commercial
       plants with incremental improvements in the technol-
       ogy as developed.

These stages of development are all designed to reduce
the technical risks and uncertainties associated with a given
system. Such risks include product yield when influenced by
scaleup, subsystem performance, materials selection, production of undesirable byproducts, and all operating and maintenance (O & M) costs. Reduction of those risks can be
accomplished only by increasing the knowledge base associated
with the technology. Increasing that knowledge base results
in progressing along the technological learning curve.
    Recognizing that there is a technology learning curve
implies recognition that there is a failure rate associated
with new technologies. This failure rate is high, and it is
a result of the complexity and fragility of technological
innovation [31]. This failure rate is one cause of the RADR.
    One partial demonstration of this approach to managing
technical risk is illustrated by the process contingency
requirement associated with the capital cost estimate of a
given technology. The Engineering Societies Commission on
Energy [10] posited an approximate process contingency cost
learning curve shown in Table I. Process contingency,
according to ESCOE, is used to cover uncertainties in un-
proven technology equipment design, performance, and cost.
It is added to the project contingency normally associated
with a capital cost estimate (usually 5-15%). This contingency may, in turn, impact the estimated annual maintenance

TABLE I.   The Capital Cost Learning Curve for New Technologies, Expressed as Contingency Estimates

Development stage	Contingency estimate[a]		
	Process cotingency (%)	*Project contingency (%)*	*Total contingency (%)*
Concept (Bench scale)	50	5	55
Process development unit	25	5	30
Pilot plant	15	5	20
Demonstration plant	10	5	15
Pioneer plant	5	5	10
Mature plant	0	5	5

[a]*Contingency is a % of direct capital costs in any total capital cost estimate.*

Source: ESCOE [10].

costs associated with a project due to the relationship between capital and maintenance costs.

Given the three critical stages of technology development identified by Swabb [34] and a project contingency of 5%, then a total contingency for the three stages of technology development are as follows: (1) pilot plant, 20%; (2) pioneer plant, 10%; and mature technology plant, 5%. Assuming that these estimated contingencies are a definitive measure of risk, there is a halving of technological risk at each stage of technology development.

2. The Cost of Capital Learning Curve. Just as there are technical uncertainties associated with new technologies, there are financial risks and uncertainties associated with such new processes. Investors, like engineers, are risk-averse. They, too, seek to reduce risk through increased knowledge. Alternatively they seek increased rewards in the form of large expected cash flows when they take large risks. This is the principle underlying the entire venture capital industry (see Rubel and Novotny [30]). This also is the principle underlying the new form of insurance, technical performance insurance (see Bailey [1]). This management of financial uncertainty through knowledge or assurances gives

rise to a cost of capital learning curve for each technology.
This cost of capital learning curve is another basis for the
risk adjusted discount rate (RADR) curve.

This application of discount rate learning curves to
technology analysis is not new, having previously been
applied to such areas as coal conversion technology evaluation [37]. What is required in this application is a techno-
logically based financial learning curve appropriate for the
FPI.

B. *Approaching Technology Analysis Through Risk Adjusted
    Discount Rate Curves*

While technology or project risk adjustment is frequently
proposed, little data exists concerning how such risk adjust-
ments should be made. Brown [5] for example, identified the
key elements in risk analysis and proposed an end point result of a 30% after tax discount rate for risky investments;
however, he proposed no analytical technique for achieving
the 30% result. Similarly PAMCO offered a discount rate curve
for the development of solvent refined coal, with no explanation of the method for reducing cost of capital values as a
function of experience [26].

In the absence of a previously published theoretical construct for making RADR calculations, the approach taken here
is to define a general RADR for manufacturing industries.
This generalized RADR is based upon empirical data concerning
marketplace behavior. This general curve is then followed by
a specific RADR curve for the FPI. The FPI specific curve is
based upon an adaptation of one theory proposed by Gordon and
Halpern [13] for estimating the cost of equity capital associated with a nontraded division or firm.

II. A GENERAL MANUFACTURING INDUSTRY RISK ADJUSTED
     DISCOUNT RATE (COST OF CAPITAL) CURVE

Three stages of technology development have been identified above as being critical: (1) the pilot plant, (2) the
pioneer plant, and (3) the mature technology plant. Of these
the discount rate for the mature technology is readily calculated using the techniques discussed in Chapter III. Values
for determining the discount rate for the pioneer plant are
more elusive, however.

Brealy and Myers [4] posit a general RADR curve appropriate for determining the pioneer plant related value. This
discount rate curve is as follows: speculative venture, 30%;

new product, 20%; expansion of existing business, 15% (company cost of capital); and cost improvement in a known technology, 10%. What is critical is the doubling of the discount rate between the normal company cost of capital and the speculative venture.

The doubling principle, also posited by ESCOE for engineering work, emerges here in the comparison of discount rates for speculative ventures and mature technologies. This RADR is quite similar to the one posited by Bierman and Smidt [2], and it is not unlike the principle proposed by Brown [5] in discussing the approaches taken by third party financial institutions to new ventures. Because the doubling principle or guideline is not infrequently proposed, it is essential to test its utility in more detail.

A.   The Generalized Cost of Capital for Pilot and Pioneer Plants

In general the cost of capital or discount rate for pilot and pioneer plants can be derived from two sources: (1) the rate of return expected by venture capitalists, and (2) the rate of return achieved by previous innovations that have entered the marketplace. Both sources of information are examined below.

1.   The Cost of Venture Capital. Venture capitalists typically invest in a project at the pilot plant stage if possible (see Geary [12]). Their expected rate of return, according to Levine [20] ranges from 35 to 60%. This high expected after tax internal rate of return reflects the fact that venture capitalists take equity positions in new high risk projects, and attempt to increase the value of their investment through an increase in the value of the firm's stock. This increase in stock value commonly occurs when the firm "goes public." This increase in value usually is taxed at capital gains rates, rather than ordinary income rates.

The range of 35 to 60% is consistent with the values posited by Hambrecht [16], however it is considerably narrower than the 15 to 100% range of expected after tax returns put forward by Danforth [8].

The 35 to 60% rate of return range appears reasonable for the purposes of this text. This range is quite broad. The breadth of this range, in itself, reflects the uncertainty associated with investments in pilot plants. Many, if not most, technologies that achieve pilot plant status never become commercial practice [31].

2. Returns on Past Commercialized Innovations. Rates of return on past innovations provide an indication of hurdle rates associated with pioneer plants, for they provide a basis for evaluating investor expectations. Further such rates of return provide the best measure of pioneer plant financial requirements, because they relate only to innovations that have gone from the pilot plant stage to commercial practice.

Robert R. Nathan Associates [28] performed one highly useful analysis of internal rates of return achieved by 20 innovations that were commercialized, 18 of which were in non-regulated private industry. Nathan Associates estimated the pretax rates of return for these innovations, establishing mean values plus upper and lower bounds.

It is significant that the innovations discussed by Nathan Associates came from a diversity of industries including mining, oil refining, agribusiness, machine tools, electronics, metals recycling, pharmaceuticals, chemicals, and telecommunications. The innovations included such major breakthroughs as numerical controls on machine tools and automobile shredding for metals recycling. Internal rates of return were uniformly distributed throughout the industries surveyed, except in the cases drawn from regulated public utilities. Such regulated public utility cases, however, are not germain to this text.

The unregulated industry cases provide perhaps the best insight into the general cost of capital or discount rate associated with pioneer plants. However those data relating to nonregulated firms must be converted from the pretax basis shown by Nathan Associates to an after tax basis. The after tax basis is necessary in order to make the Nathan Associates values consistent with the discount rates previously discussed in this text. Such a conversion can be made by assuming a mean corporate income tax rate of 40% [41], and a range of income tax rates from 35 to 45%. On this basis the after tax rates of return on innovation can be calculated, and such values are shown in Table II.

B. The General Mature Technology Discount Rate

The mature technology discount rate is readily calculated using the Weighted Average Cost of Capital (WACC) method and the CAPM cost of equity model. The assumptions and resulting calculations for a mature technology discount rate are shown in Table III.

In establishing the range of mature technology discount rates only the risk parameters (β_j and the k_d value) have been varied. The range could be somewhat broader than

*TABLE II. Estimated After-Tax Internal Rates of
Return on Innovations Commercialized by Nonregulated
Industries*

Case	Mean Internal Rate of Return (%)	Range of Internal Rate of Return values (%)
Median	25	23-27
Low	17	16-19
High	30	28-33

Source: Based on Robert R. Nathan Associates [28].

*TABLE III.   Calculation of a Mature Technology Discount
Rate for Manufacturing Industry*

Assumptions

(1) β_j *= 1.0 (0.65-1.35)*

(2) θ *= 0.33*

(3) Pretax k_d *= 11.7% (11.1-12.4%)*

(4) Tax rate = 40%

(5) K_{rf} *= 8.0%*

Results

Case	Discount rate
Mean	13.6%
Low	11.4%
High	15.9%

presented when firms have risk terms outside the range
employed (Louisiana Pacific, for example, currently has a
β_j of 1.4). For 95% of the firms in the marketplace, however, a cost of capital range of 11 to 16% is a reasonable
representation. The 11% cost of capital reflects discount
rates for very stable, highly risk averse industries. The
16% discount rate reflects conditions in more growth
oriented or riskier businesses.

It is significant to note that the current forest products industry discount rate of almost 15% is near the upper
limit of expected values. Perhaps the FPI can be considered
to be somewhat more risky, in general, than the average
industry.

C. *The General Risk Adjusted Discount Rate Curve*

The data presented above provide the essential values for
a generic RADR curve. This generic RADR curve can then be
used as a guideline for a forest industry specific RADR curve.

  1. Defining the Generic RADR Curve. The general outlines
of the generic RADR curve are summarized in Table IV. These
data suggest that, indeed, there is an approximate halving of
the discount rate as a technology passes from one major development stage to the next. This approximate halving of the
discount rate reflects increased marketplace or investor
acceptance. This phenomonon is depicted, graphically, in
Fig. 2.
  The halving phenomenon in the discount rate is significant as a tool for assessing new technologies. Equally significant is the phenomenon that the range of appropriate discount rates also shrinks as the technology stage is advanced.
This range related phenomenon is shown in Table V. The
shrinkage in the range of appropriate discount rates, on the
high side (upside risk), moves by a factor of 2.4-2.5. The
shrinkage in downside risk or uncertainty moves less
uniformly.
  These data, then, illustrate that the marketplace has
generated a general risk adjusted discount rate curve, or

*TABLE IV. An estimated General Risk Adjusted Discount
   Rate Curve for Manufacturing Industry*

	Nominal risk adjusted discount rate	
Technology stage	*Base value (%)*	*Range (%)*
Pilot plant	*48*	*35-60*
Pioneer plant	*25*	*17-30*
Mature plant	*14*	*11-16*

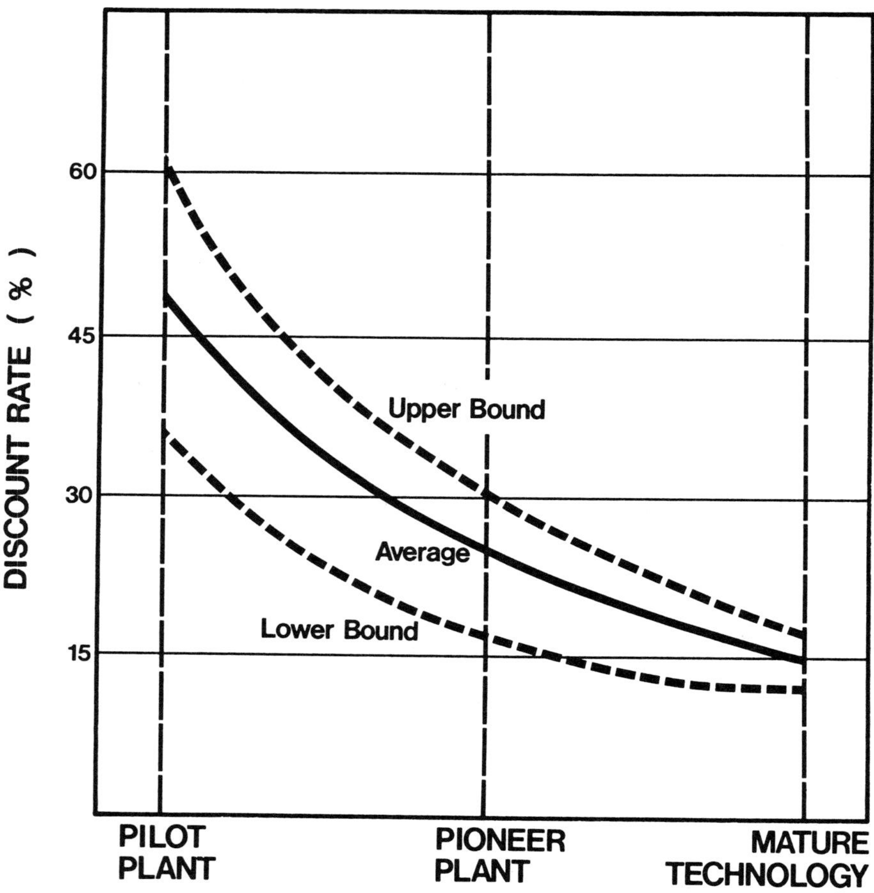

FIGURE 2. A general risk adjusted discount rate curve for manufacturing industry based on marketplace behavior.

family of curves. Such values are essential as corporations evaluate new technologies designed to expand their share of given markets, or to enter new markets.

These data can be supported, further, by insurance premiums charged by INA and other insurance firms when insuring the performance of new technologies. INA and others do insure that specific plants will meet certain design performance levels (usually defined as product outputs), and will insure them to the extent that lenders are paid if the plant fails to perform as designed. Premiums are based upon technical risk assessments performed by the insurance carrier. Higher levels of risk (e.g., plants closer to the pioneer

*TABLE V.   Variation in the Risk Adjusted Discount Rate
As a Function of Technology Stage*

Technology stage	Risk adjusted discount rate variation		
	Base value	Variation about base value	
Pilot plant	48%	+12%	−13%
Pioneer plant	25%	+ 5%	− 8%
Mature plant	14%	+ 2%	− 3%

or pilot plant stage of development) require higher premiums
[1]. Such premiums must be paid for out of the operating
income of the plant. As such they increase the revenue and
cash flow requirements of the facility and accomplish the
same result as the RADR.

   2. Using the RADR Curves. A potential investor/analyst
must select an appropriate discount rate for analysis from
the RADR curves. That discount rate must be chosen based
upon the stage of technology development. That discount rate
is then used for NPV and pricing analysis, or as a hurdle
rate for IRR evaluation
   In the case of major improvements to existing technologies, the full pioneer plant discount rate may not be
required. Rather the pioneer plant discount rate can be
applied to that portion of the investment that is at unusual
risk. Thus, the RADR for significant process modifications
might be treated as:

$$RADR_a = RADR_p(E) + MDR(1-E) \qquad\qquad (4-1)$$

Where $RADR_a$ is the risk adjusted discount rate for a major
advance in an existing technology, $RADR_p$ is the overall
pioneer plant discount rate, MDR is the mature industry discount rate, and E is the proportion of the investment at risk.
   The above discussion provides general guidelines for the
RADR approach to new investments. Given the 15% mature discount rate for the FPI, it is now essential to construct risk
adjusted discount rates specific to this industry.

III. RISK ADJUSTED DISCOUNT RATES FOR THE FOREST PRODUCTS INDUSTRY

The general discount rates and RADR curves developed above are based largely upon overall marketplace behavior. Such data do not exist specifically for Forest Industry investments carrying unusual technical risk. An alternative method for estimating the RADR curve for one industry exists, however. This method is based upon adapting the theory for estimating the discount rate of a division of a firm, or the discount rate for a nontraded firm, as developed by Gordon and Halpern [13]. This approach, while more theoretical than that employed in Section II of this chapter, does rely on case studies for data development. In using case studies, it is tied to industry behavior also. Thus it is employed here. The general theory is examined first, followed by a discussion of the adaptations necessary for its application to innovation. Following that discussion, specific FPI innovations are considered in order to develop a RADR curve for this specific industry.

A. *The Divisional Cost of Capital Theory*

Gordon and Halpern [13] addressed the problem of calculating the cost of capital for a division of a publicly traded firm in the *Journal of Finance*. The problem in such a case is the lack of data concerning the market value of equity and the market assessment of systematic risk associated with the division. Such problems also are associated with a new technology investment within an existing firm.

The Gordon and Halpern theory is based upon valuing a stock by the CAPM model discussed in Chapter III. It recognizes explicitly, however, that the CAPM and the Dividend Growth model yield cost of equity capital values that do not differ significantly.

The theory proposed by Gordon and Halpern states that there is a statistic "c_j" which is based upon expected corporate earnings. This "c_j" statistic is inextricably related to β_j by the following expression:

$$\beta_j = \lambda_0 + \lambda_1 \, c_j \tag{4-2}$$

Where β_j is the overall beta of the firm, λ_0 is the first constant, and λ_1 is the second constant. Gordon and Halpern have developed two sets of values for λ_0 and λ_1 as shown in Table VI. The regression values shown in Table VI were developed by regressing the growth rate of a sample of firms

TABLE VI.   *Values Used for Constants in the Divisional Cost of Capital Model*

Constant	Regression value	Implicit value
λ_o	.278	.564
λ_1	.584	.258

Source: Gordon and Halpern [13].

against the growth rate of the "market portfolio" [13]. The implicit values presented in Table VI were developed by regressing calculated c_j values against β_j values, and solving for the two constants, considered to be universal for U. S. firms [13]. These constants provide mathematical relationships between "c_j" and β_j which are shown in Fig. 3. Figure 3 presents calculated values using expression (4-2).

It is useful to note that the equation demonstrates that a firm with a c_j value of 0.0 will have a β_j of 0.564 based upon regression values, or 0.278 when using implicit values. Gordon and Halpern posit the explanation that other elements of systematic risk are included in the β_j term.

Following the development of c_j statistics, Gordon and Halpern compared the two measures of systematic risk, β_j and c_j. The Spearman rank correlation of the two measures was 0.72, which is highly significant. Gordon and Halpern also tested expression (4-2) and the related constants by applying them to a sample of 49 firms in both the manufacturing and public utility sectors, and found a high degree of reliability in this approach. Consequently it appears attractive as a base for adaptation here.

After developing expression (4-2), Gordon and Halpern developed the following generic equation for estimating the β of the division with respect to the firm as a whole:

$$\beta_{js} = \frac{\lambda_o + \lambda_1 c_{js}}{\lambda_o + \lambda_1 c_j} \tag{4-3}$$

Where β_{js} is the beta of the division with respect to the firm, and c_{js} is the expected earnings of the division. By substitution, then, Gordon and Halpern state:

$$k_{js} = k_{rf} + \frac{\lambda_o + \lambda_1 c_{js}}{\lambda_o + \lambda_1 c_j} (k_j - k_{rf}) \tag{4-4}$$

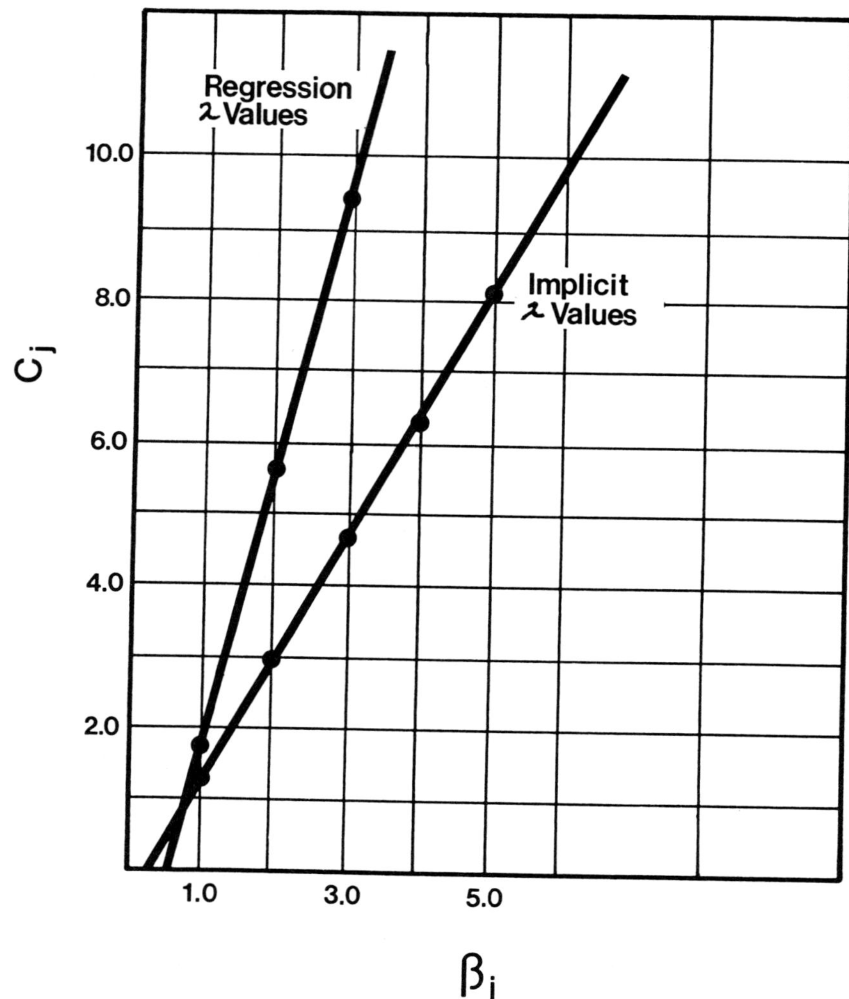

FIGURE 3. *The mathematical relationships between the "c_j" statistics and the β_j statistics based upon the equations of Gordon and Halpern [13].*

Where k_{js} is the cost of capital for the division, k_{rf} is the cost of risk free money, and k_j is the cost of capital for the firm as a whole. The cost of capital for the firm is substituted for the cost of capital in the market, as β_{js} is risk relative to β_j.

The c_j statistic for the FPI can be calculated directly from expression (4-2), using a β_j of 1.07. In such calculations, c_j is as follows: regression value - 1.36; and

implicit value - 1.96 (1.07 = .278 + .584c; alternatively,
1.07 = .564 + .258c). These values aid in providing the
range in cost of capital estimates for divisions of FPI
firms. Given these c_j values, expressions (4-3) and (4-4)
can be solved. To do so for new innovations, however, requires some adaptations of the existing theory.

B. *Adaptation of the Divisional Theory to Accommodate Technological Risk Assessment*

Three assumptions must be made if this theory is to be
applied to the calculation of discount rates for new innovations. These assumptions relate to the applicability of the
theory to the problem, and to the substitution of various
proxy values.

With respect to the applicability of this theory, it is
assumed that the Gordon and Halpern construct is as applicable to individual plants and projects as it is to entire
divisions. That assumption, based upon the Brealy and Myers
project market line [4], is somewhat limiting in that it
implicitly states that each plant must be treated as an independent profit center. The profit center assumption is con-
servative from a financial viewpoint, and it reflects a significant risk-averse posture. However the profit center
position is consistent with the behavior of many large, mature
corporations.

The substitutions into the Gordon and Halpern expressions, as required by this use of the theory, are as follows:

(1) capacity additions, or in their absence production
    values, are considered to be reasonable proxies for
    expected earnings, and

(2) β_u (the unleveraged beta) is considered to be as
    useful for analysis as β_j.

With respect to the use of physical production capacity
as a substitute for earnings, capacity is added by a firm to
garner expected earnings and cash flows. In performing
retrospective case studies, it depicts the cash flow expectations of a firm reasonably accurately. Physical production
is a less accurate estimator, however it can be used in the
absence of more appropriate data for the first decade of any
innovation.

With respect to using the unleveraged beta values, most
new technologies have been financed by equity more frequently
than a combination of debt and equity. Further, in the FPI,
it has been shown that there is little difference in the

weighted average cost of capital and the unleveraged cost of
equity capital. Consequently the model should operate ade-
quately on both WACC and β_u based discount rates.

C. *Refinement of the Divisional Theory Applied to Forest
    Industry Investments in Innovations*

Investing in any technological innovations carries the
following risks: (1) the risk of losing the capital invested in the innovation; (2) the risk of losing cash flows
that could have been earned by employing the capital in
other investments; and (3) to greater or lesser extents, the
risk of losing cash flows from other portions of the plant
where the investment is made, and which are dependent upon
the technology investment.

The first two types of risk are obvious in nature. The
third type of risk identified requires more explanation. This
type of risk can be considered in terms of reversibility of
the innovation. Consider, for example, two sawmill innovations: (1) the computer controlled, scanner guided headrig;
and (2) the microwave dry kiln. The new headrig involves
using laser lines to illustrate the best cutting approach to
the sawyer. Once the sawyer sees the cut as identified by
the laser line, the cut is then made with conventional equipment. Had the laser system failed, mills installing it could
simply have removed or ignored it, and then could have re-
verted to conventional operation. The basic bandmill or other
cutting device would have remained essentially whole. The
cash flows of the sawmill would have remained largely intact.
If, however, the sawmill owner had invested in a microwave
dry kiln and it had failed, the cash flows of the plant could
be protected only by removing the microwave device and re-
placing it with another investment such as a conventional dry
kiln. Not only would the original investment be lost, but
additional investments would be required and revenues would
be lost during the reversion process if the market being
served demanded kiln dried lumber.

The principle of innovation reversibility as outlined
above is closely related to downside risk minimization. In
principle the sawmill is more exposed to downside risk in the
microwave drying case than in the laser assisted headrig case.
In the microwave case, then, the RADR has to be higher than
in the laser case, based upon the principle of reversibility.
In the case of the new process pioneer plant, the total investment is irreversible. In the case of an innovation
within an existing technology (e.g., the sawmill), reversibility is measured by degree (see also equation 4-1).

The concept of reversibility is in line with the observation by Steele [31] that most innovations are improvements to existing processes rather than developments of totally new alternative technologies. Steele [31] states that improvements to existing technologies are more readily commercialized because they are less risky, can be adopted more quickly, and have more certain cash flow benefits. All such conditions argue for reducing the risk adjustment to the discount rate as a function of the type of innovation.

Reversibility, or the extent to which the innovation puts an entire plant at risk, influences the hurdle rate that must be achieved by the innovation, or the discount rate associated with that innovation. Such returns or RADRs are reflected in capacity addition (or production) statistics, particularly in the early years. The higher risk (largely irreversible) innovations typically will be installed by only a few firms initially, and will be installed only as a minor portion of total industry capacity. As such high risk investments demonstrate a capacity to generate exceptional cash flows, they will be adopted rapidly by other firms seeking to remain competitive. Lower risk, reversible innovations typically are installed by several firms and at significant capacity levels even at the beginning of their commercialization. Therefore the rates of capacity addition or production growth are lower for these innovations despite rapid rates of technology acceptance.[1]

Rates of technology acceptance, as measured by production capacity addition or by initial product production, provide the means for determining the "c_j" and "c_{js}" values used in expression (4-4). It is, therefore, the c statistics that are influenced directly by reversibility.

The logic behind this approach to calculation of the RADR can be summarized as follows:

(1) the highest risk innovations must generate the most favorable cash flows;

---

[1]*The growth rate difference can be shown by the following equation: $A(1.b) = C = 2A(1.d)$. In this expression, if $t>0$, $b>d$.*

The larger initial investment in the reversible innovation follows the same behavioral principle as the one exhibited in horseracing where more people will bet on the favorite in a horse race. Once the long shot horse wins 2 or 3 races, it becomes a favorite. Similarly more people will place a \$2.00 bet than a \$5.00 bet.

(2) such high cash flows, for past innovations, will be
    reflected in high rates of capacity addition relative
    to total industry rates of capacity growth; and

(3) as innovations are perceived as being less risky,
    they will attract larger populations of initial in-
    vestors, hence more firms will invest initially and
    the associated rates of technology growth will be
    proportionately lower (other things being equal).

Given these assumptions, the Gordon-Halpern [13] divisional cost of capital model can be applied to past FPI innovations in order to develop pioneer plant and process modification discount rates. These past innovations will provide a
basis for establishing a RADR curve specific to the Forest
Products Industry.

D.   *Case Studies in FPI Innovation Used in Calculating
     an Industry RADR*

Four primary case studies are presented below as a means
for calculating a RADR curve for the FPI: (1) thermomechanical pulping (TMP); (2) the use of continuous digesters in
kraft pulping; (3) the manufacture of particleboard; and
(4) the production of pulp chips by sawmills. These four
case studies cover a 23 year period of innovation, 1950-1973.
They focus heavily upon efforts to utilize mill residuals
most economically, since that has been a major thrust of the
industry. Each case is presented in terms of its major technoeconomic features, and the resulting derived discount rate.
In addition to the primary cases identified above, two
innovation cases have been analyzed at a more cursory level:
(1) the Kamyr displacement pulp bleaching system, and (2) the
Lamb-Cargate wet cell. These two secondary cases are used to
support the conclusions developed from the four primary innovation case studies. In addition to the examination of primary innovations, this chapter also considers the second
generation of specific new technologies.
In presenting the primary case studies, a time frame must
be established. This time frame is assumed to be the lower
bound of the asset depreciation range (ADR) cited by the
Internal Revenue Service in depreciation and plant life publications. Thus TMP pulping systems have a projected evaluation period or tax life of 11 years while particleboard
plants have an evaluation period of 8 years. Second generation technologies also have been examined in these case
studies, and in such cases the analytical period is extended
to the midpoint of the ADR range.

1. Thermomechanical Pulping. TMP pulping, as discussed in Chapter I, is now considered state-of-the-art in mechanical pulping. This technology offers numerous features of economic significance including: (1) it can use residual chips, rather than pulp logs, as furnish; (2) it has a yield of 95%, comparable to other mechanical pulping processes; and (3) it produces a mechanical pulp with the highest percentage of long fibers, and therefore TMP produces a pulp of sufficient strength to replace much of the reinforcing chemical (kraft) pulp in the production of newsprint [38].

Selected TMP pulp quality characteristics are compared to stone groundwood pulp characteristics in Table VII. What is not shown in Table VII is the fact that TMP pulp is not only stronger than stone groundwood pulp, but it is also stronger than refiner mechanical pulp (RMP), and is therefore more useful than both other mechanical pulps in the production of many products [45]. The properties shown in Table VII are those characteristics that permit TMP pulp to displace higher cost bleached kraft pulp in newsprint manufacture [9]. They also permit TMP pulps to be used in the manufacture of light-weight coated magazine papers, catalogue papers, wall papers, business forms, sanitary products, and a host of other applications [32, 45].

TMP pulp was invented by A. Asplund of Defibrator Corp., in 1939. The basic principle involves preheating the chips under pressure prior to refining. Preheating softens the wood and results in the liberation of more long fibers during

TABLE VII. A Comparison of Characteristics of Thermomechanical and Stone Groundwood Pulps

Parameter	Pulp type	
	Thermo-mechanical	*Stone groundwood*
Burst Index (kPa·m²/g)	*2.3*	*1.4*
Tear Index (mN·m²/g)	*9.0*	*4.1*
Breaking length (meters)	*3300*	*2000*
Bulk (cm³/g)	*3.0*	*2.6*
Opacity (%)	*96.0*	*97.0*
Brightness (%)	*56.0*	*58.0*
Shives (Somerville)	*0.05*	*0.10*

Sources: West [45].

the refining stage. In early process attempts TMP pulps suf-
fered from brown stain. The lignin, softened during preheating, was smeared over (and ultimately hardened on) the individual fibers [11]. Commercial introduction of the TMP process
came with the solution to this staining problem.

The first commercial TMP mill was a 1968 Swedish installation [32]. The first mill in North America was installed by
Publishers Paper Co., Newberg, Oregon, in 1973 [33]. St.
Anne Paper Co. (Abitibi) of Beaupre, Quebec, installed the
second North American TMP line in 1974 [14]. The growth of
TMP pulping since 1973 has been dramatic. This growth resulted from the utility of the pulps and the low capital cost
of TMP pulping systems [38]. It has come despite the high
electricity requirements of TMP pulping—typically 1800-
2700 kWh/O.D. ton of pulp depending upon the degree of
refining [38].

Capacity growth of TMP pulping since 1973 is shown in
Table VIII. With respect to Table VIII, it is interesting to
note that, in the year of introduction, only one plant was
built. Today there are 23 installations in 13 states including Alabama, Georgia, Maine, Minnesota, Texas, Oregon, and
Washington. The high energy cost associated with the TMP
process apparently has not influenced plant location. This
location independence is reinforced by the analysis of TMP
pulping economics in the Southeast performed by Chiang [6].
Further, while the first plant installed had a capacity of
200 ton/day, the largest current unit is a 1,450 ton/day
plant in Longview, Washington, owned by Northern Pacific

TABLE VIII. Capacity Growth Rates for Thermomechanical
   Pulping in the United States and in North America
   (Values in tons × 10^3)

	Annual capacity	
Year	United States	North America
1973	60	60
1977	740	1210
1980	1200	2240
1984	2800	4100
Annual capacity growth rate (%)	42	47

Sources: Post's 1984 [27] and previous editions.

Paper Co. (NORPAC) and shown in Fig. 4. NORPAC is a subsidiary of the Weyerhaeuser Co. and Jujo Paper Co., Ltd. of
Japan [27].

The growth of TMP was sufficiently strong that it induced
a response from the threatened technology: groundwood pulping. After the introduction of TMP, Tampella and others
countered with a major innovation in the groundwood process:
pressurized groundwood pulping. PGW produces a pulp that is
superior to conventional groundwood pulp. Further, it is
more economical to produce PGW than groundwood pulp [25]. PGW
retains the basic limitation, however, that it can not use
residual chips.

To place the growth of TMP pulping into perspective, it
is necessary to examine the growth of total mechanical pulping capacity in the U.S. In 1973 total U.S. mechanical pulping capacity was about 5 million tons/yr [38], while in
1984, mechanical pulping capacity was 6.1 million tons/yr
[27]. TMP pulping, then, began with 1.2% of the mechanical
pulping capacity and within 11 years had captured 45% of that
market. TMP capacity growth over this period was 42%/yr
while total mechanical pulping growth was 2%/yr.

FIGURE 4. Thermomechanical pulp refiners at the NORPAC
pulp mill in Longview, Washington. (Photo courtesy of
Weyerhaeuser Co.)

These capacity growth factors are used to determine the c_{js} factor. Because the c_{js} factor as presented by Gordon and Halpern is calculated on a nominal dollar basis, some rate of inflation must be employed in the ultimate calculations. During the period 1973-1984, the overall rate of inflation as measured by the implicit price deflator series, was 7.7%. Therefore a c_{js} factor can be approximated at 5.[1] Given this c_{js} factor, the pioneer plant discount rate can be calculated. First the β_{js} term is calculated:[2]

$$\beta_{js} = \frac{.564 + .258(5)}{.564 + .258\ (2.0)} = 1.72$$

$$\beta_{js} = \frac{.278 + .584(5)}{.278 + .584\ (1.4)} = 2.92$$

Once the β term exists, the discount rate can be calculated either on an nominal or real basis. The rate is determined as follows:

$$k_{js} = 0.029 + 1.72(0.097 - 0.029) = 0.15 \quad \text{(real)}$$

$$k_{js} = 0.029 + 2.92(0.097 - 0.029) = 0.23 \quad \text{(real)}$$

Then

$$k_{js} = 1.15 \times 1.05 - 1 = 0.21 \quad \text{(nominal)}$$

$$k_{js} = 1.23 \times 1.05 - 1 = 0.29 \quad \text{(nominal)}$$

The 9.7% shown above is the real unleveraged DR for the FPI as shown in Chapter III. The use of the TMP case, then, yields a nominal pioneer plant discount rate of 21-29%, or a real rate of 15-23%.

For sensitivity analysis purposes, c_{js} values of 4 and 6 were also used. The resulting estimated real and nominal discount rates are shown in Table IX. Based upon the values in Table IX, the nominal discount rate for FPI pioneer plants with new processes is estimated at 21-29%. This range is in the middle of the generalized RADR curve shown previously.

---

[1]*This calculation is as follows:*
$((1.42 \times 1.077)-1)/((1.02 \times 1.077)-1) = 5.34$.
[2]*The values for c_j used in the denominator, 1.4 and 2, have been derived by back calculation of c_j terms for the Forest Products Industry, using a β_j value of 1.07 as shown in Chapter III.*

TABLE IX.   *Risk Adjusted Discount Rates for the Forest
Products Industry Based Upon Thermomechanical
Pulping*

Parameter	Case		
	Low	Base	High
C_{js} *factor derived from capacity growth rate*	4	5	6
Resulting β_{js}	1.5-2.4	1.7-2.9	2.0-3.5
Real discount rate range (%)	13-19	15-23	17-27
Real discount rate midpoint (%)	16	19	22
Nominal discount rate range (%)[a]	19-25	20-29	22-33
Nominal discount rate midpoint (%)[a]	22	25	28

[a]*Inflation = 5%*

2. *The Continuous Digester for Kraft Pulping.* Kraft
pulping was invented in 1884 by a Swedish chemist. It was
introduced to the U.S. in 1907 [21]. By the 1950s the kraft
process had become the dominant pulping method employed in
the U.S. and the free world. Its dominance was based upon
its applicability to a wide variety of wood species, its
ability to readily accomplish chemical recovery as part of
the overall process, and its ability to produce high strength
pulps [38]. Kraft pulping benefitted from major innovations
during its growth years, particularly the development of the
Tomlinson Recovery Boiler to achieve both chemical and energy
recovery from spent pulping liquors [40].

The kraft process, as invented, was a batch process. The
heart of the system, the digester, was periodically charged
with chips and pulping liquor, and then operated. After the
cook, the digester was unloaded. In 1938 Kamyr A.B. of
Sweden began experimentation with a continuous digester
process [44]. Throughout World War II experiments proceeded
at the Kamyr 5 ton/day pilot plant. In 1948 the large scale
50 ton/day pilot plant was installed and operated for final
process development [43]. The first commercial unit, a 165
ton/day system [17], was installed, in Sweden, in 1952.

The modern continuous digester, as shown in Fig. 5,
offers advantages to kraft pulping in the areas of energy

FIGURE 5. The modern Kamyr continuous digester for kraft pulping. (Photo courtesy of Leaf River Forest Products.)

conservation, product uniformity, and overall process efficiency [21]. It gained acceptance throughout the industry
rapidly. The first units were introduced into the U.S. in
1958, and growth since that time has been substantial. This
growth in continuous digester capacity for kraft pulping in
the U.S. is shown in Table X.

By 1971, continuous digesters accounted for some 30% of
kraft pulping capacity in the U.S. By 1984 these systems
commanded 40% of the kraft pulping capacity. During this
period of time the digesters were continuously improved as
Fig. 5 infers. By 1980 the U.S. achieved a dominant position in continuous digester kraft pulping, as is shown in
Table XI. These growth statistics must be put into perspective with regard to the type of innovation characterizing the
continuous digester—the irreversible innovation. The Kamyr
digester is irreversible in that failure of the continuous
digester at a plant would have required its removal, coupled
with reinvestment in batch digesters. Without removal and
reinvestment the entire mill would be unusable. Investing in
the continuous digester, at the early stages of commercial
introduction, could jeopardize the cash flows of the entire
pulp mill. Consequently the discount rates calculated from
this case study are irreversible innovation (existing
process) discount rates.

TABLE X.   Continuous Digester Capacity as a Percent
of Total Capacity in the U.S. Kraft Pulping Industry
(Tons/yr $\times$ 10^3)

Year	Continuous digester capacity	Total kraft pulping capacity	Continuous digester capacity share (%)
1958	210	12,730	1.7
1971	8,100[a]	27,410[c]	29.6
1984	17,300[b]	43,240[d]	40.0

[a] Annual capacity growth rate, 1958-1971 = 32.5%

[b] Annual capacity growth rate, 1971-1984 = 6.0%

[c] Annual capacity growth rate, 1971-1984 = 6.1%

[d] Annual capacity growth rate, 1971-1984 = 3.5%

Source: Post's 1984 [27] and previous editions.

*TABLE XI.   Continuous Digester Capacity by Country in
    1980   (Value in tons × 10^3)*

Country	Annual capacity	Percent of total continuous digestor capacity
United States of America	*16,500*	*34.9*
Canada	*7,600*	*16.1*
Sweden	*4,800*	*10.1*
Soviet Union	*4,200*	*8.9*
Japan	*3,600*	*7.6*
Finland	*2,500*	*5.3*
Brazil	*1,800*	*3.8*
All others	*6,300*	*13.3*
Total	*47,300*	*100.0*

Source: Virkola 1983 [43].

Risk adjusted discount rates have been calculated from this case study. From 1958–1971 the annual rate of increase in continuous digester capacity was 32.4% in the U.S. During that same period of time the annual rate of increase in total kraft pulping capacity in the U.S. was 6.1%. During those 13 years the rate of inflation in the U.S. was 2.9%. These data yield a c_{js} factor for the continuous digester innovation of 4. The resulting discount rate for the FPI, along with sensitivity analyses, is shown in Table XII. The base case RADR is 16%. The base case RADR is 16% (real) or 22% (nominal).

3. *The Particleboard Manufacturing Case Study.* While the Masonite ^{tm} process has existed since nearly the turn of the 20th Century, dry particleboard processing was not invented until 1941, and was not commercialized until after World War II [23]. Early plants were installed between 1947 and 1951 in Wilton, New Hampshire, in the midwest, and in Longview, Washington [23]. The thrust of dry particleboard processing is to produce a product from residues that can be substituted for plywood and/or lumber in a variety of applications.
    Reliable capacity data are unavailable for this discount rate analysis and, therefore, production data are used as an

TABLE XII. Risk Adjusted Discount Rates for the Forest Products Industry Based on Continuous Digester Capacity

Parameter	Case		
	Low	*Base*	*High*
c_{js} factor	3	4	5
Resulting β_{js} factor	1.2-1.9	1.5-2.4	1.7-2.9
Real discount rate range (%)	11-16	13-19	15-23
Real discount rate midpoint (%)	13	16	19
Nominal discount rate range (%)[a]	17-22	19-25	20-29
Nominal discount rate midpoint (%)[a]	19	22	25

[a] Inflation = 5%

admittedly less precise measure. This substitution is made recognizing that production is a less precise measure of earnings expectations than capacity additions. Production data for particleboard are shown in Table XIII. From Table XIII it is apparent that, for the period 1950-1958, production of conventional wood products was decreasing at a rate of 1.2%/yr while production of particleboard was increasing at a rate of 36%/yr.

Because production is employed as a measuring device, producer prices also are employed in order to convert physical quantities into earnings. From 1950 to 1958 the producer price of lumber increased at a rate of 0.4%/yr while plywood prices decreased by 1.1%/yr [42]. During this period the rate of inflation was 2.9%/yr. During this period stumpage values were rising at a real rate of 0-2%/yr depending upon species and grade. These data suggest an overall earnings decline from lumber and wood products of 5-6%/yr.

The data presented above result in a calculated c_{js} factor of about 5 for particleboard manufacturing during its initial commercialization. Consequently the values in Table IX hold for particleboard as well as TMP pulp.

4. The Pulp Chip Case Study. The final primary case study is the production of pulp chips from sawmill residues,

TABLE XIII. Production of Particleboard and Solid Wood Products in the U.S. (In tons $\times 10^6$)

Year	Production	
	Particleboard	Solid wood products
1950	0.04	44.7
1952	0.06	44.5
1954	0.09	43.6
1956	0.21	46.3
1958	0.47	40.7
1960	0.50	40.7
1962	0.77	42.2
1964	1.20	47.4
1966	1.48	48.5
1968	2.12	48.8
1970	2.63	46.9

Source: USDA Forest Service, 1981 [42].

a practice that began on the west coast in about 1950 [46]. Producing pulp chips from sawmill residue was an innovation of enormous importance. It changed the sawmill from an operation that converted logs into one product, lumber, into a plant that made several products from harvested timber. At the same time it reduced the cost of pulp mill furnish. Conversion of the sawmill from a single product technology to a multiple product technology meant that this production system could serve multiple markets, as shown in Fig. 6. Thus this innovation reduced, to some extent, the cyclical influences of the housing industry on the economics of sawmill operation.

In order to accomplish the production of pulp chips at sawmills, lumber producers had to add debarkers, slab chippers, and chip bins to their processes. Prior to the production of chips, sawmills fed barky logs directly to the headrig [46]. The addition of pulp chip production not only improved product recovery from the log, but also increased saw life and decreased maintenance costs associated with lumber production [46].

FIGURE 6. Pulp chips have become a major commodity and cash flow generator for sawmills.

Despite the tremendous significance of this innovation, the production of pulp chips from sawmill residues exhibited a high degree of reversibility. Had the debarkers and chippers failed, barky logs again could have been fed to the headrigs using prior practices. Reversibility meant that investments to produce pulp chips did not threaten the cash flows from lumber production. This reversibility is in stark contrast to the conditions that existed for investors in the continuous digester.

Because this innovation was largely a west coast phenomenon during the 1950s, it has been analyzed with west coast data, recognizing that the industry was highly regionally oriented at that time. This analysis again employs production data in the absence of available, capacity information. From 1950 to 1958 pulp chip production from sawmill residues increased from 1.0 million tons/yr to 3.3 million tons/yr [15]. This is an annual rate of increase in production of 16.1%. At the same time west coast lumber production increased from 18.6 billion bd ft/yr to 19.7 billion bd ft/yr, at a rate of 1%/yr [42]. These data, along with inflation and producer prices, result in a calculated c_{js} factor of 3 for the innovation converting sawmill residues into pulp chips. Such a c_{js} factor results in a nominal RADR of 19% or a real RADR of 13%, or for reversible innovations added to existing processes (see Table XIV).

5. *Secondary Case Studies.* Two secondary case studies were given cursory examination: (1) displacement bleaching, and (2) gasification-combustion. These case studies support the data developed above.

Displacement bleaching, like the continuous digester and the diffusion washer, was invented by Kamyr, A.B. of Karlstad, Sweden. Displacement bleaching allows up to five bleaching stages (e.g., C-E-D-E-D) to be performed in a single tower or reactor vessel. As such it reduces the capital and operating costs of pulp bleaching. Diffusion bleaching, for example, reduces bleach related electrical energy costs by up to 45% [18].

The displacement bleaching system was first introduced in 1972, when 110 ton/day of capacity was installed. By 1982, 9540 tons/day of displacement bleaching capacity had been installed throughout the world. The largest single unit has 1220 ton/day of capacity [19]. The growth rate of displacement bleaching in the free world was 56%/yr during its first 10 years. That growth rate was comparable to continuous

TABLE XIV. Risk Adjusted Discount Rates for the Forest Products Industry Based Upon the Production of Pulp Chips From Sawmill Residues

	Case		
Parameter	*Low*	*Base*	*High*
C_{js} *factor*	2	3	4
Resulting β_{js}	1.0-1.3	1.2-1.9	1.5-2.4
Real discount rate range (%)	10-12	11-16	13-19
Real discount rate midpoint (%)	11	13	16
Nominal discount rate range (%)[a]	16-18	17-22	19-25
Nominal discount rate midpoint (%)	17	19	22

[a]*Inflation = 5%*

digester capacity growth from 1952–1962 on an inflation adjusted basis. Consequently the growth of displacement bleaching supports the irreversible innovation RADR.

The second support case study is the Lamb-Cargate close coupled gasifier-combustor or wet cell. This unit contains both a gasifier and a low Btu gas combustion unit (see Chapter VII for a technical description). The first wet cell, shown in Fig. 7, was installed in 1978 at Plateau sawmills, Engen, British Columbia. Recent installations include units for composite products furnish drying (see Fig. 8) and boiler retrofits (e.g., Proctor and Gamble, Baltimore, Maryland).

The first wet cell had a capacity of 25×10^6 Btu/hr. At the end of 1984, some 20 wet cells were installed with a combined capacity of 650×10^6 Btu/hr. While the period of analysis here is sufficiently short that rigorous analysis is impossible, and while the same period experienced the Iranian revolution and consequent wood energy growth [36, 39], such a growth rate (64%/yr) supports the pioneer plant RADR.

6. Second Generation Innovations and Resulting RADRs.
Of final interest is the second generation of any technology or innovation. Data have been shown for both the continuous kraft digester (Table X) and particleboard manufacturing (Table XIII). These data provide insights into the rate of risk reduction as a technology moves from the first to the second generation.

In the continuous digester case, capacity increased at an annual rate of 6.0% between 1971 and 1984. During this period, total kraft pulping capacity increased at a rate of 3.6%/yr. The resulting c_{js} factor for continuous digester kraft pulping is slightly above 2. The discount rate associated with this innovation approached the rate associated with mature investments.

In particleboard, the second generation plants emerged in about 1960 [23]. From 1960 to 1970 particleboard production increased by 18%/yr while total board product production increased by 9%/yr. These data lead to a calculated c_{js} factor somewhat greater than 2, but closer to 2 than 3. These data suggest that the technical risk associated with the innovation had been largely eliminated. Second generation systems exhibit a discount rate approximately 1% greater than the mature technology discount rate.

E. A Risk Adjusted Discount Rate Curve for the FPI

The data provided above offer some indication of a risk adjusted discount rate curve for the FPI. This RADR function is summarized in Table XV, and is plotted against the general

FIGURE 7. *The first commercial Lamb-Cargate wet cell, installed at the Plateau Sawmills, Engen, British Columbia. (Photo courtesy of Lamb-Cargate Industries.)*

FIGURE 8. A recent installation of Lamb-Cargate wet cells at a composite products manufacturing facility. (Photo courtesy of lamb-Cargate Industries.)

TABLE XV. An Estimated Risk Adjusted Discount Rate Curve for the Forest Products Industry

Technology status	Mean risk adjusted discount rate (%)	
	Nominal	Real
New process pioneer plant	25	19
Irreversible innovation	22	16
Reversible innovation	19	13
Second generation technology	16	11
Mature technology	15	10

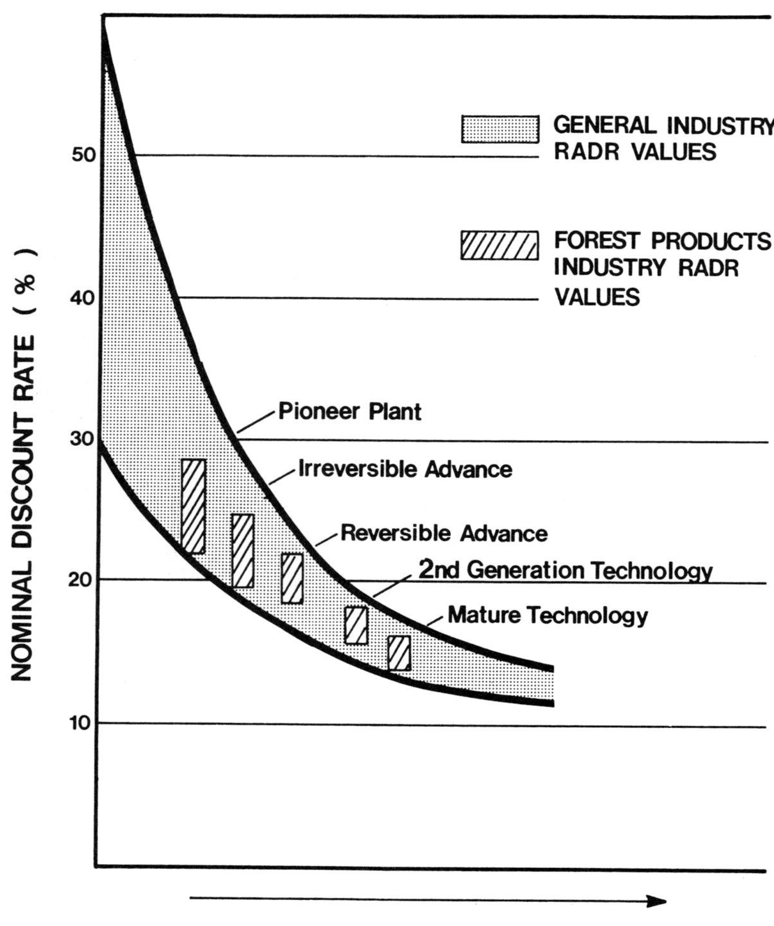

FIGURE 9. *A risk adjusted discount rate curve derived for the Forest Products Industry.*

industry RADR in Fig. 9. This RADR function indicates that there is a steady progression in risk reduction, leading to a lower discount rate, depending upon the nature of the innovation being commercialized. Further, this function includes the fact that second generation (incremental) advances in innovations have discount rates only slightly above those costs of capital associated with mature technologies.

IV. CONCLUSION

The marketplace employs RADR values. These rates can be observed from the yields to maturity on bonds as a function on bond rating, from the beta values associated with stocks, from the behavior of the venture capital industry, from the current practices of insuring technological performances on new systems, and from numerous other data. These RADR values are critical in analyzing new investments.

Techniques do exist to calculate RADR curves both generally and for a specific industry. When employed these techniques yield a risk related function well within the bounds of an overall industry RADR curve, as is shown for the FPI relationship to the overall curve in Fig. 9.

Given the relationship shown in Fig. 9, it is now possible to evaluate specific proposed technologies for the forest products industry. Such proposed technologies exist in conventional areas such as pulp production and sawmilling. They exist, also, in such areas as the production of fuels and energy, chemicals, and foodstuffs. While the number of proposed new technologies makes exhaustive treatment impossible, selected advances are considered in subsequent chapters using the model contained in Chapter II, and the discount rates as calculated in Chapter III, and in the above sections.

REFERENCES

1. Bailey, G. J. 1984. Insurance for Energy Risks. *In* "Energy Technology XI: Applications and Economics," Proc. Government Institutes, Inc., Washington, D.C., 164-166.

2. Bierman, H., and Smidt, S. 1971. "The Capital Budgeting Decision," 3rd Ed. MacMillan Co., New York.

3. Bisio, A., and Gastwirt, L. 1978. Turning Research and Development into Profits. Amacon, Division of American Management Association, New York.

4. Brealy, N., and Myers, S. 1981. "Principles of Corporate Finance." McGraw Hill, New York.

5. Brown, D. L. 1984. Unpublished remarks by the President of Time Energy Systems, Inc. at Energy Technology XI Conference, Washington, D.C., March 19-21.

6. Chiang, T. I. 1978. Economics of Thermomechanical
    Market Pulpmills in the Southeastern Part of the United
    States. *Forest Products Journal 28*(7):18-23.

7. Cox, L. A. 1974. Transfer of Science and Technology
    in Successful Innovation. *Forest Products Journal 24*(9):
    44-48.

8. Danforth, P. D. 1971. Venture Capital Techniques . . .
    and Tomorrow. *In* "How to Raise and Invest Venture
    Capital" (S. M. Rubel and E. G. Novotny, eds.). Presi-
    dents Publishing House, New York.

9. Dillen, S., and Soderlund, U. 1975. Will TMP Replace
    Groundwood or Chemical Pulp in Newsprint. *In* TAPPI
    International Mechanical Pulping Conference Proc., San
    Francisco, California, June 16-20.

10. Energy Societies Commission on Energy. 1979. Guidelines
    for Economic Evaluation of Coal Conversion Processes.
    ESCOE, Washington, D.C.

11. Evans, J. C. W. 1978. Exclusive Worldwide TMP Survey.
    Pulp and Paper 52(6):101-110.

12. Geary, G. S. 1971. Types of Ventures Financed. *In* "How
    to Raise and Invest Venture Capital" (S. M. Rubel and
    E. G. Novotny, eds.). Presidents Publishing House, New
    York.

13. Gordon, M. J., and Halpern, P. J. 1974. Cost of Capital
    for a Division of a Firm. *Journal of Finance 29*(4):1153-
    1163.

14. Goucher, W. A. 1975. Thermomechanical Pulping - Abitibi
    Paper Company, Ltd., Beaupre Division. *In* TAPPI Inter-
    national Mechanical Pulping Conference Proc., San
    Francisco, California, June 16-20.

15. Grantham, J. B. 1974. Status of Timber Utilization on
    the West Coast. U.S.D.A. Forest Service, Portland,
    Oregon.

16. Hambrecht, W. R. 1971. Forming a Venture Capital Firm.
    In "How to Raise and Invest Venture Capital" (S. M.
    Rubel and E. G. Novotny, eds.). Presidents Publishing
    House, New York.

17. Kamyr A. B. 1981. Continuous Cooking and Washing Sys-
    tems. Bulletin No. 201-JR4. Kamyr, Inc., Glens Falls,
    New York.

18. Kamyr A. B. 1981. Diffusion Washers and Displacement Bleaching Systems. Bulletin No. 500-a. Kamyr, Inc., Glens Galls, New York.

19. Kamyr. 1984. Displacement Bleaching: Six Systems Sold in One Year (Advertisement). *Pulp and Paper* 58(1):99.

20. Levine, J. 1983. Money for the Asking. *Venture*, June, 34-40.

21. Libby, E. C. 1962. "Pulp and Paper Science and Technology," Vol. 1. McGraw-Hill, New York.

22. Lindsay, R., and Sametz, A. W. 1963. "Financial Management: An Analytical Approach." Richard D. Irwin, Inc., Homewood, Illinois.

23. Maloney, T. M. 1977. "Modern Particleboard and Dry-Process Fibreboard Manufacturing." Miller Freeman Publications, San Francisco.

24. Myers, S. C. 1968. Procedures for Capital Budgeting Under Uncertainty. *Industrial Management Review*, Spring.

25. Paljakka, V. J., Malkki, R., Savia, R., and Sinkko, T. 1984. PGW Benefits LWC Paper Production in Two years of Use at Finnish Mill. *Pulp and Paper* 58(6):73-76. Also see Evans, C. W. 1980. Pressurized Process for Groundwood Production Making Healthy Progress. *Pulp and Paper* 54(6):76-78.

26. Pittsburgh and Midway Coal Mining Co. 1979. "Conceptual Commercial Plant Economic Analysis," Vol. 5. For U.S. Department of Energy, Denver, Colorado.

27. Post's. 1984. "Pulp and Paper Directory." Miller Freeman Publications, San Francisco. Also see Post's for years from 1958-1983.

28. Robert R. Nathan Associates, Inc. 1978. Net Rates of Return on Innovations: Final Report. RRNA, Washington, D.C. (For National Science Foundation.)

29. Robichek, A. A., and Myers, S. C. 1966. Conceptual Problems in the Use of Risk-Adjusted Discount Rates. *Journal of Finance* 21(9):727-730.

30. Rubel, S. M., and Novotny, E. G. 1971. "How to Raise and Invest Venture Capital." Presidents Publishing House, New York.

31. Steele, L. 1983. Managers' Misconceptions About Technology. *Harvard Business Review* 61(6):133-140.

32. Strauss, R. W. 1978. Status of Thermomechanical Pulp (TMP) and Its Impact on the Southern Paper Industry. *In* "Complete Tree Utilization of Southern Pine," Proc. (C. W. McMillin, ed.). Forest Products Research Society, Madison, Wisconsin.

33. Strom, B. J. 1975. Thermo-Mechanical Pulping at Publishers-Newberg. *In* TAPPI International Mechanical Pulping Conference Proc., San Francisco, California, June 16-20.

34. Swabb, L. E. 1978. Liquid Fuels From Coal: From R & D to an Industry. *Science 199*(4329):619-622.

35. Tillman, D. A. 1978. "Wood As An Energy Resource." Academic Press, New York.

36. Tillman, D. A., Rossi, A. J., and Kitto, W. D. 1981. "Wood Combustion: Principles, Processes, and Economics." Academic Press, New York.

37. Tillman, D. A. 1981. Technical Risk as a Cost Factor in Coal Liquefaction: Some Financial Considerations. Preprints, Fuels Div., American Chemical Society, 26(3): 141-154.

38. Tillman, D. A., Rossi, A. J., and Simmons, S. O. 1982. Wood: Its Present and Potential Uses. Envirosphere Co., Bellevue, Washington. (For the U.S. Congress, Office of Technology Assessment.)

39. Tillman, D. A. 1983. Making the Best Energy Use of Wood. *In* "Progress in Biomass Conversion," Vol. 4. Academic Press, New York.

40. Tomlinson, G. H., II. 1976. Black Liquor Recovery—An Historical Note. *In* "Forum on Kraft Recovery Alternatives," Proc. Institute of Paper Chemistry, Appleton, Wisconsin.

41. U.S. Department of Commerce. 1984. Statistical Abstract. U.S. Government Printing Office, Washington, D.C.

42. U.S.D.A. Forest Service. 1981. U.S. Timber Production, Trade, Consumption, and Price Statistics 1950-1980. Washington, D.C. Miscellaneous Publication #1408.

43. Virkola, N. E. 1983. Puumassan Valmistus: Suomen Paperi-insinoorien Yhdistyksen oppi- ja kasikirja II.

44. Wenzl, H. F. J. 1970. "The Chemical Technology of Wood" Academic Press, New York.

45. West, W. B. 1979. Mechanical Pulping as of Today.
    TAPPI 62(6):19-21.

46. Williston, E. M. 1984. Personal communication with
    author, Jan. 3. Also see: Williston, E. M. 1981.
    Small Log Sawmills. Miller Freeman Publications, San
    Francisco; and Williston, E. M. 1976. Lumber Manufac-
    turing: The Design and Operation of Sawmills and Planer
    Mills. Miller Freeman Publications, San Francisco.

Chapter V

ADVANCED PULPING PROCESSES

I. INTRODUCTION

The manufacturing of pulp, and the products derived from
pulp, is the most significant economic sector of the forest
products industry (FPI). Pulping accounts for the most tons
of products manufactured, the highest sales volume, and the
most substantial capital investments in the FPI (see Chapter
I). Consequently any examination of the advanced technologies in the FPI must focus first on an evaluation of wood
pulping.
There are a vast number of pulping processes extant today.
Further, a large number of additional potential innovations
in wood pulping could be adopted by the end of this century.
Consideration of each present and potential wood pulping
process is beyond the scope of this text. As a consequence,
this chapter considers the advanced pulping systems by
(1) discussing the current state-of-the-art in pulping, (2)
surveying some of the systems being proposed to advance wood
pulping, and then (3) examining two proposed systems in some
detail. These detailed reviews are designed to illustrate
the use of the discounted cash flow models and techniques
discussed in Chapter II, and using the discount rates developed in Chapters III and IV. At the same time these technology case studies are intended to highlight directions of
technological advances being proposed today.

II. THE STATE-OF-THE-ART IN WOOD PULPING

Chemical delignification is the dominant pulping process
being used today. Numerous chemical pulping technologies
exist including kraft, acid sulfite, alkaline sulfite, neutral sulfite, bisulfite, and soda pulping. Some chemical
pulping systems may be combined with mechanical pulping
(e.g., neutral sulfite semichemical or the cold soda

pulping). Unlike mechanical pulping, where yields range
from 85-95%, chemical pulping systems have yields in the 35-
55% range depending upon the species being pulped, the severity of the pulping conditions, and the extent of bleaching.
     Kraft pulping, as depicted in Fig. 1, is the dominant
chemical pulping technology largely due to its early success
in pulping virtually any wood species and because chemical
recovery and reuse is easily accomplished. Kraft pulps also
exhibit superior strength properties when compared to most
other chemical or mechanical pulps. Because it is the dominant chemical pulping technology, it is the focus of this
section.

A. *The Principles of Kraft Pulping*

     In kraft pulping wood is subjected to a liquor containing
sodium sulfide (Na_2S) and sodium hydroxide ($NaOH$). Sodium is

FIGURE 1. *The state-of-the-art in bleached kraft pulping is represented by the Leaf River Pulp Mill. This
view highlights the bleach plant and the Kamyr digester.
(Photo courtesy of Leaf River Forest Products.)*

the base of this liquor, with the active agents being the hydroxyl (OH^-) and hydrosulfide (HS^-) ions [32]. Typical pulping conditions for the kraft process are summarized in Table I [1]. The kraft cook, occurring in either a continuous or batch digester, liberates the individual long cellulosic fibers (D.P. = 8,000 to 10,000) and delignifies them. Delignification makes the fibers more flexible, giving the pulp sheet more contact points between the fibers, hence higher strength.

The kraft pulp leaving the digester contains some residual lignin, and it is the bleaching process that removes the final quantities of that aromatic natural high polymer. Bleaching is accomplished with chlorine (C), alkali extraction (E), chlorine dioxide (D), hypochlorite (H), oxygen (O), ozone (Z), hydrogen peroxide (P), and other similar compounds. Perhaps the most common bleaching sequence is C-E-C-E-D.

Kraft pulp leaving the digester is separated from the spent pulping liquor by washing. The spent liquor, containing 12-15% solids (lignin, extractives, some hemicelluloses, and spent pulping chemicals) is then concentrated in evaporators and concentrators, and burned in the recovery boiler for chemical and energy recovery. The chemicals removed from the boiler in the form of a green liquor smelt then enter the chemical regeneration circuit and the process starts again [28].

TABLE I.   Typical Kraft Pulping Parameters

Parameters	Value
Time to maximum temperature	*1.0 - 2.5 hrs.*
Time at maximum temperature	*1.0 - 2.0 hrs.*
Cooking temperature	*330 - 350°F (165 - 175°C)*
Liquor/wood ratio	*3.0 - 4.0*
Chemical charge (% active alkali on bone dry wood)	*12 - 18%*
Sulfidity (% Na_2O on AA)	*15 - 35%*

Source: Aho, 1983 [1].

B. *The Process of Kraft Pulping*

As the summary of kraft pulping principles suggests, this
process involves not only the digestion of wood chips, but
also numerous unit operations for continuous use of the pulping chemicals and for byproduct (including energy) recovery.
The process can be viewed through a series of simplified
flowsheets concerning (1) wood and fiber flows, (2) chemical
flows, and (3) energy flows.

1. Wood and Fiber Flows. Wood and fiber flows in the
kraft process are depicted in Fig. 2. This figures is exclusive of the bleach plant. It assumes that the process has a
yield of about 47% (42% on a barky log basis), and produces
1,500 lb black liquor solids/A.D. ton of pulp (1,800 lb pulp
O.D.) as shown in Babcock and Wilcox [5], and Hurley [18].
With respect to Fig. 2, it is useful to note that energy
flows are not proportional to product yields. If one assumes
that 1 O.D. ton of wood contains 17×10^6 Btu (8,500 Btu/lb)
and that one ton of spent liquor dry solids contain $13.2 \times
10^6$ Btu [5], then the approximate distribution of energy
flows is as follows: bark to boiler, 1.8×10^6 Btu; pulp,
6.0×10^6 Btu; black liquor (to boiler), 9.2×10^6 Btu; and
total, 17.0×10^6 Btu. About 65% of the energy flows are
contained in the 58% of the wood and bark that does not end
up as pulp.

2. Chemical Flows in the Pulp Mill. Chemical flows are
far more complex than wood flows in the kraft mill as Fig. 3
shows. Fig. 3 is a schematic flowsheet of an unbleached
kraft mill. The chemical recovery operations are in white
boxes and the nonchemical recovery operations are in cross
hatched boxes shown in Fig. 3.
What becomes apparent from Fig. 3 is the fact that the
recovery boiler is the real heart of the pulp mill. It is
the point at which spent chemicals begin their return to a
useful state. Also it becomes apparent that the chemical
recovery system is a high profile target for major improvements in the system.

3. Energy Flows in the Kraft Mill. Energy flows are the
third critical area of concern. Steam is the major form of
energy used, and the steam demands of the unbleached kraft
mill are summarized in Table II. These steam demands do not
include the thermal energy required for the bleach plant or
the paper machine, but they do include steam consumed by a
pulp dryer. On the basis of Table II one can calculate that
1 oven dry ton (O.D. ton) of unbleached kraft pulp requires
the expenditure of 16.76×10^6 Btu in the form of steam,

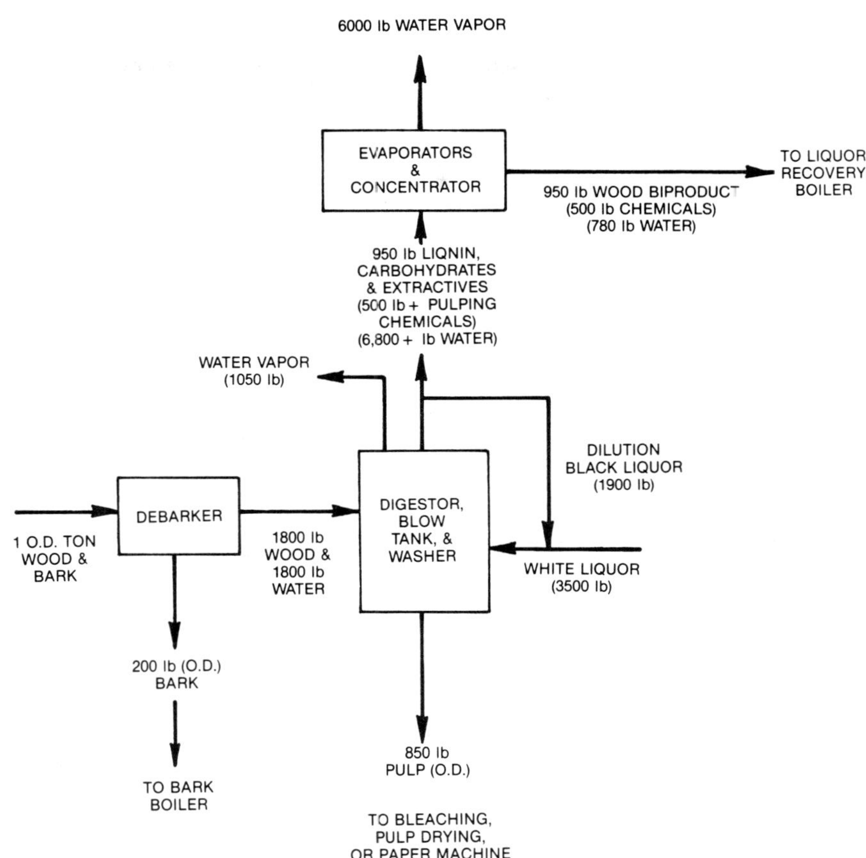

FIGURE 2. Typical wood flows in an unbleached kraft pulp mill. (Adapted from Hurley [18]).

given a 47.5% yield of pulp from wood furnish [18]. Electricity consumption is about 950 kWh/O.D. ton of pulp. Oil consumption for the lime kiln is about 2.0×10^6 Btu/O.D. ton of pulp [1, 18]. These energy requirements are in addition to the steam consumed.

As a practical matter the unbleached kraft mill supplies much of its energy requirement, and can be nearly 100% steam self-sufficient. However cogeneration typically is employed such that the mill generates most of its own power in addition to steam as illustrated by Fig. 4. A modern unbleached kraft mill might be 90% steam self-sufficient and over 85% electricity self-sufficient depending upon such site specific

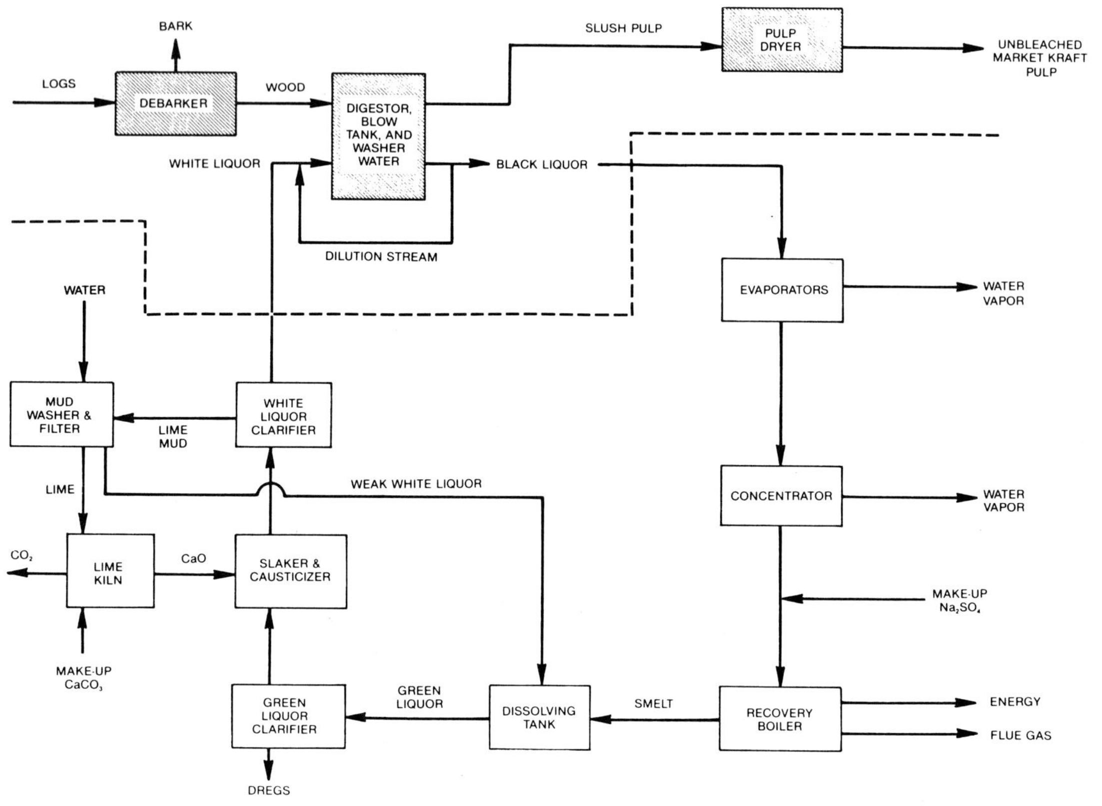

FIGURE 3. A schematic flow sheet of an unbleached kraft pulp mill highlighting the chemical flows in the mill. (Adapted from Minor [28]).

TABLE II. Energy Flows in the Unbleached Kraft Mill   (Basis: 1 O.D. Ton Pulp)

Point of use	Steam demand (lbs)	Steam pressure (psig)	Steam enthalpy (Btu/lb)	Steam heat flow (Btu × 10⁶)	Condensate return (lbs)	Condensate heat flow (Btu × 10⁶)
Digesters	3,593	150[a]	1,196	4.297	2,156	0.539
Air heater	1,049	150[a]	1,196	1.255	997	0.332
Liquor heater	72	150[a]	1,196	0.086	–	–
Shatter jet	12	150[a]	1,196	0.014	–	–
Pulp dryer	5,000	150[a]	1,196	5.980	3,000	0.750
Miscellaneous	1,000	150[a]	1,196	1.196	600	0.150
Evaporation	3,158	65[b]	1,183	3.736	3,000	0.819
Miscellaneous	2,000	65[b]	1,183	2.366	1,200	0.300
Soot blowing	600	400	1,205	0.723	–	–
Total	16,484	N/A	N/A	19.653	10,953	2,890

Lime kiln: 2.0 × 10⁶ Btu as oil or gas.

Electricity: 950 kWh

[a] First stage cogeneration turbine exhaust.

[b] Second stage cogeneration turbine exhaust.

Source: Hurley [18] updated to accommodate a 10% increase in energy conservation.

FIGURE 4. The Kattua, Finland, kraft pulp mill, one of the first locations of an advanced circulating fluidized bed combustor designed to burn wood waste and peat, illustrates the near energy self-sufficiency of kraft pulping. (Photo courtesy of Pyropower.)

conditions as fiber source (cord wood vs. residual chips), specific process configuration employed, cogeneration system design (including pressure and temperature of the high pressure steam generated), and other factors as shown in Table III. (See Hurley [13], Tillman [34], Tillman [35].) Virtually self-sufficient bleached kraft mills (90% steam; 100% electricity) also have been configured [3].

C. The Costs of Kraft Pulping

The principles and processes shown above can be translated into capital, operating, and maintenance charges associated with kraft pulping. Such costs are detailed below. For costing purposes a 1000 ton/day mill is used.

1. Capital Costs. The conceptual capital costs of 1000 ton/day green field unbleached and bleached kraft mills are shown in Table IV, in disaggregated fashion. It is useful to note that the wood and pulp related facilities represent $187 million dollars or 42.5% of the capital costs in the case of the unbleached kraft mill, and $275 million or 50% of the capital costs in the case of the bleached kraft mill. Chemical recovery systems for these units cost nearly $140 million, 31% of the capital costs of the unbleached kraft mill or 25% of the capital costs of the bleached kraft mill. It is useful to note that the recovery boiler is the largest single capital investment in the pulp mill. These data demonstrate that pulping and chemical recovery are logical targets for economic improvement.

2. Operating and Maintenance Costs. Operating and maintenance costs for the units also can be estimated, and these costs are presented in Table V. From Table V the apparent factor is the preponderance of costs in such areas as wood, labor, energy, and chemicals, and the desirability of cost savings in such areas.

3. Levelized Costs. The data presented above can be converted into levelized costs per ton of bleached and unbleached pulp, as is shown in Table VI. Since this book assumes marginal analysis (analysis of the next increment of capacity), the values presented in Table VI are employed for alternative technology investment analysis.

4. Cost/Price Squeeze. The data above provide a sound basis for comparative analysis. They fail to elucidate the severe cost/price squeeze now facing the chemical pulping

TABLE III.   Heat Balance of Unbleached Kraft Pulp Mill
(Basis: 1 O.D. Ton of Pulp)

Assumptions

Pulp yield	*47.5%*
Pulp furnish	*40.0% residual chips* *60.0% cordwood*
Black liquor	*3,397 lb/O.D. ton BLS (3057 lb/A.D. ton)* *6,600 Btu/lb*
Bark	*9,000 Btu/lb (O.D.)*
Boiler efficiencies	*64.0% black liquor* *70.0% hog fuel*
Cogeneration turbine	*Throttle steam = 1250 psig/950°F* *First stage (η=.75) exhaust = 150 psig* *Second stage (η=.80) exhaust = 65 psig*
Generator efficiency	*95.0%*
Water to desuperheaters	*80°F (h = 48)*
Steam flows	*10,726 lb @ 150 psig to system* *5,158 lb @ 65 psig to system* *9,920 lb @ 150 psig from boiler[a]* *4,936 lb @ 65 psig from boiler[a]*
Water flows	*10,953 lb @ 264 Btu/lb to system* *3,903 lb @ 48 Btu/lb to system*

[a]*The rest comes from desuperheaters (7.6% of 150 psig steam and 4.3% of 65 psig steam).*

Values

(1) Fuel to hog fuel boiler

$$2000 \text{ lb pulp} \times \frac{lb\ wood}{.475\ lb\ pulp} \times \frac{.6\ lb\ cordwood}{lb\ wood}$$

$$\times \frac{.11\ lb\ bark}{lb\ cordwood} \times \frac{9000\ Btu}{lb\ bark} - 2.5 \times 10^6\ Btu$$

(Continued)

TABLE III. (Continued)

---

(2) Steam from hog fuel boiler

$$2.5 \times 10^6 \; Btu \times .7 \times \frac{lb}{(1468 - 207)Btu} = 1388 \; lbs \; (9.3\%)$$

(3) Fuel to black liquor boiler

$$1 \; O.D. \; ton \; pulp \quad \frac{3397 \; lb \; BLS}{O.D. \; ton \; pulp} \times \frac{6600 \; Btu}{lb \; BLS} =$$

$$22.42 \times 10^6 \; Btu$$

(4) Steam from black liquor boiler

$$22.42 \times 10^6 Btu \times .64 \times \frac{lb}{(1468-207)Btu} = 11,379 \; lb \; (76.6\%)$$

(5) Steam from power boiler

$$14,856 \; lb \; demanded - (11,379 + 1,388) \; lb \; supplied =$$

$$2,089 \; lb \; required \; (16.5\%)$$

(6) Fuel to power boiler (e.g., coal)

$$2,089 \; lb \times \frac{(1468-207)Btu}{lb} \times \frac{1 \; Btu \; in}{0.85 \; Btu \; out} = 3.10 \times 10^6 Btu$$

(7) Power from first turbine stage

$$14,856 \; lb \; steam \times \frac{kWh}{14.354 \; lb} \times .75 = 776 \; kWh$$

(8) Power from second turbine stage

$$4,936 \; lb \; steam \times \frac{kWh}{48.68 \; lb} \times 0.8 = 81 \; kWh$$

(9) Net power to plant

$$(776 \; kWh + 81 \; kWh) \times .95 = 814 \; kWh \; (95.8\%)$$

(Continued)

TABLE III.   (Continued)

---

(10) Purchased power

(950-814) kWh = 136 kWh (14.3%)

(11) Purchased oil for lime kiln

2×10^6 *Btu*

Sources: Hurley [18]; Tillman [34].

industry; and it is this squeeze that is critical to the need
for accelerated innovation.

Aho [1] demonstrated that the cost of producing bleached
kraft pulp in a southern mill was about 370/air dried tonne
(ADMT) or $350/air dried ton (ADT) in 1982. These values
translate into about $410/ADMT or $380/ADT in 1984. The cost
components were as follows: feedstock (wood), 30.6%; labor,
15.2%; energy, 13.2%; chemicals, 11.3%; other direct costs,
13.4%; and indirect costs and overhead, 16.3%. These data
are based upon mills of varying ages, from relatively new or
rebuilt plants to fully depreciated operations.

Aho [2] also demonstrated that the costs of critical factors of production are rising at faster rates than the prices
of pulp, paper, and allied products. Factors of production
where costs are rising faster than product prices include
labor, energy, and chemicals. Such factors include pulpwood
in the Pacific Northwest. The cost of pulpwood in the south
is rising at the same rate as the price of pulp based
products.

Operating costs are putting cost pressures on pulp mills.
At the same time the capital contribution is of serious concern. Pulp mills are said to be "too big to be small and too
small to be big." Granskog [16] demonstrated that the minimum new mill is 1000 ton/day. Most new mills are in the
1600 ton/day to 2000 ton/day range. This is a far cry from
the petrochemical concept of economies of scale. A small oil
refinery of 50,000 barrels (bbl) per day produces 7,400 tons
of product per day while the more normal refinery output
would be 200,000-300,000 bbl/day or 30,000-45,000 ton/day of
product. The petrochemical concept of economies of scale is
particularly important in light of the intermaterials competition discussed in Chapter I.

Economies of scale appear not only in capital cost but
also in labor and maintenance charges. Further, they are

*TABLE IV. Capital Cost Estimate of the 1000 Ton/Day
   Unbleached Kraft Pulp Mill (1984 Dollars)*

Cost category	Cost
Direct costs	
General facilities	23,700,000
Wood handling	40,620,000
Pulping, washing, and screening	50,780,000
Pulp drying	47,390,000
Black liquor recovery boiler	81,240,000
Solid fuels bark and power boiler	20,310,000
Cogeneration turbine and related systems	27,070,000
Recausticizing system	20,310,000
Wastewater treatment system	27,080,000
Subtotal	338,500,000
Indirect costs	
Engineering @ 10%	33,850,000
Construction management @ 10%	33,850,000
Working capital @ 5%	16,900,000
Contingency @ 5%	16,900,000
Total	440,000,000

*Source: Tillman, Rossi, and Simmons [33] as updated;
Ekono [11].*

not overcome by any apparent differences in feedstock costs.
The cost of pulpwood charged to bleached kraft pulp is about
$115/O.D. ton ($105/A.D. ton) in 1984 dollars (based upon
Aho [1]. This reflects 1984 pulpwood costs of about $56/
cord (or $53/O.D. ton) in the South, not including transpor-
tation to the mill (pulpwood cost escalated from U.S.D.A.
Forest Service [36]). It also reflects bleached kraft pulp
yields on the order of 44 to 45% (see Hirsch [17]; Tillman,
Rossi, and Simmons [33]). The cost of wood to an unbleached
mill is about $91/ADT. Oil currently costs slightly less
than $30/bbl, or about $200/ton. If one compares feedstock
costs on a $/unit of product basis, however, this near 2:1

TABLE V.   Annual Operating and Maintenance Costs for a New 1000 Ton/Day Kraft Mill   (1984 $ $\times$ 10^3)

Cost	Unbleached kraft mill	Bleached kraft mill
Wood	39,050,000	42,160,000
Labor	20,940,000	20,940,000
Chemicals	6,230,000	15,570,000
Oil (@ $30/bbl) for lime kiln	3,750,000	3,750,000
Coal (@ $2/Btu $\times$ 10^6)	2,170,000	2,170,000
Electricity (@ 6¢/kWh)	2,860,000	3,160,000
Maintenance (@ .03 $\times$ capital costs)	13,200,000	16,500,000
Supervision and overhead	22,490,000	22,490,000
Total	110,690,000	126,740,000
Cost per ton	316/ton	362/ton

Source: Aho [1] as updated.

advantage in feedstock costs is overturned. For example there are 25,000 paper bags/ton and 50,000 film plastic bags/ton [7]. Consequently the feedstock costs are $0.0046/bleached kraft paper bag or 0.0040/unbleached kraft bag and $0.0040/plastic bag.

Wood pulp based products, then, enjoy no particular economic advantages with respect to their plastic and petrochemical competitors. Rather, they may be at serious disadvantage.

D.   Improving the State-of-the-Art in Wood Pulping

If the pulping segment of the FPI is to halt the erosion of its packaging markets, and eliminate future potential competition in other markets, it must address the cost disadvantages now prevailing between it and other industries. Specifically, it must address the issues of: capital cost per ton, feedstock cost per ton, and operating cost per ton (including energy).

*TABLE VI.   Calculation of Levelized Costs for
Kraft Pulping*

	Mill type	
Cost categories	*Unbleached*	*Bleached*
Capital cost values		
Capital cost	$\$440 \times 10^6$	$\$550 \times 10^6$
Investment tax credit	*10%*	*10%*
Discount rate	*15%*	*15%*
Project life	*30 yrs*	*30 yrs*
ACRS (depreciation) life	*5 yrs[a]*	*5 yrs[a]*
Inflation	*5%/yr*	*5%/yr*
Operating and maintenance values (1984)		
Fixed operations and maintenance cost	*$56,630,000/yr*	*$59,930,000/yr*
Labor Maintenance Supervision		
Variable operations and maintenance cost	*$164/ton*	*$205/ton*
Wood Chemicals Oil and coal Electricity		
Levelized production prices		
Nominal dollars	*$700/ton*	*$850/ton*
Real dollars (1984$)	*$450/ton*	*$540/ton*

*[a]This simplified analysis ignores the fact that certain
portions of the capital investment such as buildings are
depreciated over time periods of 10-18 years.*

III. ADVANCED CONCEPTS FOR WOOD PULPING

There has been no dearth of concepts to improve the
process of wood pulping. Some of these are innovations now
being put in place. Others have been proposed and are being
evaluated. These concepts have impacted both mechanical and
chemical pulping. These concepts range from minor modifications of existing processes to alternative pulping methods.

A. Mechanical Pulping

Mechanical pulping has experienced dramatic progress in
the past two decades with the introduction and maturation of
TMP pulping, the introduction of pressurized groundwood (PGW)
pulping, the development of the Garden State newsprint
recycle process and the development of Old Corrugated Container recycling. Such innovations have improved the quality of
mechanical pulps and shifted the furnish base towards residues.

Within TMP pulping, recent improvements have been made in
the capture of waste heat generated by the process [39, 27].
Heat recovery rates (as a percent of total energy input) of
66–86% have been achieved by using TMP reject heat for air
and water heating, or for producing steam through mechanical
or thermal recompression. This recovered waste heat can be
used in space heating or in integrated mill boiler applications. TMP energy consumption levels of 2200–2700 kWh/ton
can be mitigated by such waste heat recapture systems.

B. Chemical Pulping

Chemical pulping improvements have occurred or have been
in most major areas of the pulp mill: the digester, the
recausticizing circuit, and the chemical recovery system.
Incremental improvements also have been achieved in the
bleach plant.

1. Digester Improvements. Digester related improvements have proceeded since the advent of the continuous
digester (see Fig. 5). Such improvements include the addition of anthraquinone (AQ) catalyst in kraft and soda pulping. AQ increases the pulp yield by 2–3%. The advantages
of AQ pulping are highly site specific. Catalyst recovery
remains an economic question of considerable consequence, as
it has not been accomplished to date. AQ costs $2.00/lb
(see Goel, Ayroud and Branch [14]; McDonough and Van Drunen
[26]; Aho [1]).

Figure 5. The Metsa-Botnia kraft mill in Finland,
one of the earlier locations of the continuous digester.
(Photo courtesy of Ekono, Inc.)

   2. Causticizing System Improvements. Causticizing involves making strong alkali from weak alkali. The causticizing system currently revolves around the lime kiln and
the limestone calcining process that generates calcium oxide
and carbon dioxide from calcium carbonate.

   In the area of limestone calcining, the lime kiln currently is fired with oil or natural gas. Recently, however,
sawdust has been burned in suspension as a substitute for
premium fossil fuels in the lime kiln at the Lovholmen
unbleached kraft mill of AB Svenska Skogskndustrier (ASSI).
Dried sawdust is burned along with peat at this installation
[25], shown in Fig. 6. Sawdust contains <1% ash, and therefore presents little lime contamination hazard. Experimentation also has demonstrated the ability to fire a lime kiln
with the products of combustion from a Lamb-Cargate close
coupled gasifier/combustor or wet cell. These products of
combustion can contain sufficient moisture content to exhibit
nonluminous radiative heat transfer properties. At the same
time the combustor is designed to minimize particulate carry-
over (e.g., to <0.05 gr/DSCF). These experiments were

FIGURE 6.   The Lovholmen unbleached kraft mill in Sweden, location of the peat and sawdust burning lime kiln. The biomass firing system is in the foreground of the photograph. (Photo courtesy of Ekono, Inc.)

conducted at the MacMillan Bloedel mill in Port Alberni,
British Columbia, as shown in Fig. 7. Replacement of up to
78% of the oil used to fire a lime kiln was achieved in the
Port Alberni experiments [10, 40].

The above programs involve fuel substitution within the
constraints of the existing lime kiln. More fundamental
changes have been proposed as well involving the principle of
autocausticizing, or the formation of strong alkali from weak
alkali without the addition of causticizing chemicals.

The first alternative is the direct alkaline recovery
system. Ferric oxide or titanium oxide can be used to causti-
cize spent pulping liquor from the (nonsulfur) soda pulping
process. This DARS process is based upon adding ferric oxide
to the spent pulping liquor and driving the following
reactions [41]:

*Figure 7. One of the two lime kilns fired by the
Lamb-Cargate wet cell at the MacMillan Bloedel mill
in Port Alberni, British Columbia. (Photo courtesy
of Lamb-Cargate Industries.)*

$$Na_2CO_3 + Fe_2O_3 \longrightarrow Na_2OFe_2O_3 + CO_2 \qquad\qquad (5\text{-}1)$$

$$Na_2OFe_2O_3 + H_2O \longrightarrow 2NaOH + Fe_2O_3 \qquad\qquad (5\text{-}2)$$

Ferric oxide produced by reaction (5-2) then can be used as a reactant in reaction (5-1).

Autocausticizing can be applied to kraft pulping also. In this autocausticizing system, sodium borates are used as pulping chemicals. Autocausticizing using borates or equivalent chemicals, particularly tetrasodium borate, made from boraxo (as $2Na_2O \cdot B_2O_3$) or boric acid has significant capital cost advantages, and may have operating cost benefits. It is sufficiently important that it is one of the two case studies analyzed in the next section of this chapter.

3. Chemical Recovery Boiler Improvements. The modern kraft mill chemical recovery boiler, as shown in Fig. 8, has received as much research attention as the causticizing circuit. Numerous alternatives to the conventional recovery process have been proposed. Recovery systems proposed, including those achieving pilot plant status, are the St. Regis Corp. hydropyrolysis process, the Weyerhaeuser Co. dry pyrolysis process, and the University of California pyrolysis-gasification process [8, 9, 15, 18]. These processes have not been commercialized largely because their improvements relative to the Tomlinson furnace have not overcome their technical risks.

While alternative recovery units have not been commercialized, extended surface evaporators (concentrators) have been implemented. Concentrators replace direct contact evaporators. They have improved the energy efficiency of the conventional Tomlinson recovery boiler by reducing the moisture content of the incoming kraft black liquor to about 35%. Such concentrators also control the malodorous methyl mercaptans (CH_3SH) as the methyl mercaptains are contained and are not released to the atmosphere.

Vapor recompression systems have been incorporated with the multiple effect evaporators that precede concentrators for maximized thermodynamic efficiency of the total system. The thermal efficiency of the state-of-the-art chemical recovery system has increased from about 60% to about 64% on average (for calculated values see Hurley [18], Frederick [12], and Grace [15]).

4. Pulping System Alternatives. The improvements discussed above consider unit operations within existing pulp mill types. Alternatives to existing processes also have been proposed. Such alternatives include the merging of

FIGURE 8. The modern black liquor recovery boiler at
the Leaf River Mill, with multiple effect evaporators
in the foreground. Research to date has not developed
an economically viable alternative to this system.
(Photo courtesy of Leaf River Forest Products.)

chemical and mechanical pulping systems, and pulping in
organic solvents such as methanol (MeOH).

Among the most significant innovations are those bringing
mechanical and chemical pulping together. Chemithermomechan-
ical pulping (CTMP) and chemimechanical pulping (CMP) provide
treatment of pulp chips with sodium sulfite and caustic soda
at concentrations of 2-10% prior to their introduction into
the refiners. Pulps produced from these processes have significantly higher strength than those produced by the present
mechanical processes. This strength improvement adds to the
market value of such pulps [6]. CTMP pulps may replace sulfite pulps in many applications.

CTMP is a most interesting process since it builds upon
the current state-of-the-art in mechanical pulping (TMP).
CTMP pulping has the following characteristics in addition to
producing higher strength pulps: (1) It has pulp yields of
85-92% [6] compared to typical chemical pulping yields of
45-50%, and kraft-AQ or sulfite yields no greater than 55% [1];
and (2) The CTMP process reduces the specific energy consumption per ton of pulp by up to 40% as a function of NaOH addition [24].

While the advantages of CTMP appear largely as strength,
yield, and energy consumption improvements, it is most significant that the capital investment in CTMP is about half of
that associated with kraft or sulfite pulping on an equivalent capacity basis [6,11]. Consequently this process has
strong economic advantages relative to either conventional
mechanical or chemical pulping.

Pulping in organic solvents (organisolv pulping) is a
second process innovation of note. Pulping in liquors of low
molecular weight alcohols (e.g., MeOH, EtOH, or butanol) and
water, with or without catalyst addition, promises savings in
such areas as capital and energy cost. Organisolv mills can
be configured as net energy producers [33]. Because organisolv pulping is radically different from existing pulping
processes, it is the second innovation subjected to detailed
analysis.

A host of other potential pulping innovations exist including oxygen delignification, the use of the Masonite[tm]
process for producing wood pulp, the use of soda-amine systems
[30], and a wide variety of bleaching concepts. The long
range perspective includes biological pulping (pulping aided
by fungal attack) now being researched by the U.S. Forest
Service and the Swedish Forest Products Laboratory. One
enzyme proposed for biological pulping is derived from
"white rot." It can achieve pulping and bleaching simultaneously [6]. For analytical purposes here, however, autocausticizing and organisolv pulping appear more interesting for
the remainder of this century.

IV. ECONOMIC ANALYSIS OF BORATE BASED AUTOCAUSTICIZING AND ORGANISOLV PULPING

Autocaustizing with borates and organisolv pulping are subjected to detailed economic analysis, and for the following reasons: (1) they are highly touted technologies, proposed as either a major improvement in or an alternative to the existing chemical pulping processes; (2) they represent various levels in risk; and (3) they represent the range of alternatives available for analysis (either a unit operation within a major process or a total new process).

The analyses of autocausticizing and organisolv pulping presented below recognize the interrelationships between technical and economic issues. They are presented in the following sequence: (1) process principles, (2) process overview, and (3) engineering economics.

A. *Autocausticizing with Borates*

Autocausticizing, the formation of strong alkali from weak alkali without the addition of causticizing chemicals, has considerable appeal. It permits the elimination of the recausticizing system associated with the kraft mill (see Fig. 3). Elimination of recausticizing provides for a capital cost savings of about $26,400,000 ($1.3 \times $20,310,000 as shown in Table IV) in a new 1000 ton/day mill. Elimination of recausticizing also achieves a redistribution and potential reduction of the operating costs shown in Table V. The analysis of borate based kraft pulping presented below includes evaluating the fundamentals of autocausticizing chemistry, the basic process considerations, and then the economic concerns.

1. The Chemistry of Autocausticizing. Janson [19], generally considered to be the originator of borate based autocausticizing, presents the generic chemistry of the system in the following sequence, assuming that NaA represents some sodium salt and OrgH represents alkali consuming wood components:

$$NaA + H_2O \rightleftharpoons NaOH + HA \tag{5-3}$$

$$NaOH + OrgH \rightleftharpoons OrgNa + H_2O \tag{5-4}$$

With the net reaction being

$$NaA + OrgH \rightleftharpoons HA + OrgNa \tag{5-5}$$

Oxidation in the Tomlinson recovery boiler then is represented
by

$$HA + OrgNa + XO_2 \longrightarrow Ha + 1/2\ Na_2CO_3 + YH_2O + ZCO_2 \qquad (5\text{-}6)$$

And finally if HA (an amphoteric compound) has the ability to
expel CO_2 from the Na_2CO_3:

$$HA + 1/2\ Na_2CO_3 \rightleftharpoons NaA + 1/2\ H_2O + 1/2\ CO_2 \qquad (5\text{-}7)$$

Reaction (5-7) is reversible but, with continuous removal of
H_2O and CO_2 as is accomplished in any combustor, it can be
driven almost completely to the right.

Reaction (5-7) creates NaA, the original pulping chemical. Alternatively, if HA does not survive the oxidation
process (reaction 5-6), then Ha will be transformed into the
anhydro compound, HA', and the following reaction will proceed:

$$HA' + 1/2\ Na_2CO_3 \rightleftharpoons NaA'\ 1/2\ H_2O + 1/2\ CO_2 \qquad (5\text{-}8)$$

This reaction also creates the original pulping chemical.
Reactions (5-7) and (5-8) are the keys to autocausticizing.

Numerous chemicals can be used in autocausticizing systems including borates, phosphates, aluminates, and silicates.
Of these chemicals, tetrasodium diborate is most favorable
because it provides the required alkalinity of the pulping
liquor (pH = 13-14) without presenting the scaling problems
associated with silicate based chemicals [19, 20, 21]. The
use of tetrasodium diborate simplifies the pulping process
chemistry, as is shown in Table VII. This simplification
leads to the process and economic characteristics of autocausticizing.

2. Process Considerations. The borate based kraft pulping system gains the advantage of simplicity in chemistry
without sacrificing the process characteristics associated
with conventional kraft pulping. These characteristics include pulp quality, yield, and energy consumption for
pulping.

Jansen [21] has documented the fact that borate based
pulping yields a product with the same strength properties
as conventional kraft pulping. Further, Jansen [19, 20, 21]
has shown that the borate based autocausticizing system is
applicable to both hardwoods and softwoods. Qualitatively,
then, the pulps suffer no degradation when disodium tetra-
borate is the primary pulping chemical.

The borate process exhibits unbleached pulp yields of
47-55% depending upon the species being pulped and the degree

*TABLE VII. A Comparison of Chemical Reactions Associated
With Borate Based Autocausticizing and Conventional
Kraft Pulping*

Process location	Required reactions	
	Borate-based autocausticizing	*Conventional kraft pulping*
Digester	$HOrg + Na_2HBO_3$ $\downarrow$ $NaOrg + NaH_2BO_3$	$HOrg + NaOH$ $\downarrow$ $NaOrg + H_2O$
Recovery furnace	$2NaOrg + 2NaH_2BO_3$ $\downarrow$ $Na_2CO_3 + 2NaBO_2$ $\downarrow$ $Na_4B_2O_5$	$2NaOrg$ $\downarrow$ Na_2CO_3
Mixing tank	$Na_4B_2O_5$ $\downarrow$ $2Na_2HBO_3(aq)$	Na_2CO_3 $\downarrow$ $Na_2CO_3(aq)$
Causticizing plant		$Na_2CO_3(aq) + CaO$ $\downarrow$ $2NaOH + CaCO_3$
Lime kiln		$CaCO_3$ $\downarrow$ $CaO + CO_2$

Source: Jansen, 1980 [21].

of delignification desired [20]. Borate pulping may increase
the product yield slightly relative to conventional kraft
pulping. Chemical consumption is lower with this process,
although the chemical is more expensive.

The energy consumption associated with borate pulping is
a more complex issue. While borate pulping eliminates the

lime kiln, hence the consumption of 2×10^6 Btu oil/ton of
pulp, the liquor recovery boiler is less efficient as is
shown in Table VIII. The borate based system has a liquor
recovery boiler with a thermal efficiency of 52%, compared to
the conventional black liquor boiler efficiency of 64% [3].
The possible energy gain from autocausticizing is the ability
to substitute cheap coal for expensive oil in the pulp mill
by substituting a steam generation requirement for the lime
kiln. Energy consumption values for autocausticizing are
shown in Table IX, assuming favorable conditions.

3. Economics of Autocausticizing. The economic advantage of autocausticizing is that it achieves both capital and
operating cost savings relative to conventional kraft pulping,
particularly when it is installed in a green field mill. Capital cost savings are derived from simplification of the system and elimination of the recausticizing circuit. Total
capital cost savings are on the order of $26 million for a
1000 ton/day mill.

Operating cost savings are achieved in the areas of labor
and maintenance due to process simplification. Energy expenditures are reduced not as a consequence of improved thermodynamic efficiency, but by substituting lower cost coal for
higher cost oil. While fossil fuel Btu consumption increases,
expenditures for fossil fuel decrease. Expenditures for chemicals do increase when the borate system is used. On balance,
however, there is a net decrease in operating costs as shown
in Table X.

The costs discussed above, plus the discount rate, are
necessary elements for the ultimate economic analysis of the
project. The risk adjusted discount rate (RADR) chosen for
this assessment is the reversible innovation discount rate.
In the case of failure of autocausticizing, no capital system
needs to be removed and replaced for the mill to be converted
to conventional kraft pulping. Conversion requires only the
addition of the well proven recausticizing circuit.

On that basis the engineering economy measures of net
present value and levelized production price have been calculated, and these are shown in Table XI. It is useful to note
that, on a risk adjusted basis, borate pulping is about competitive with conventional kraft pulping. The nominal net
present value (NPV) is -$1.1 million. This can be converted
into investment benefit/cost (B/C) ratio as follows:

$$\frac{\$(396 - 1.1) \times 10^6}{\$396 \times 10^6} = .997$$

TABLE VIII. Comparison of Heat Balance for Conventional and Autocausticizing Liquor Recovery Boilers

	Boiler type	
Heat loss (% of heat input)	*Conventional*	*Borate*
Sensible heat, dry flue gas	*5.1*	*5.7*
Sensible heat, combustion air moisture	*0.1*	*0.1*
Fuel hydrogen to water vapor	*6.7*	*9.5*
Evaporation of water in fuel	*10.4*	*13.7*
Sensible heat of smelt	*3.6*	*5.2*
Saltcake reduction	*1.2*	*1.2*
Heat of reaction correction	*6.5*	*6.4*
Borate dehydration	*-*	*1.7*
Autocausticizing	*-*	*1.5*
Radiation and manufacturing margin	*2.8*	*2.8*
Total losses	*36.4*	*47.9*
Boiler efficiency	*63.6 (say 64.0%)*	*52.1 (say 52.0%)*

Source: Aho, 1984 [3].

The NPV value at maturity yields a B/C ratio of 1.04, with net benefit streams being 14.9 to 17.5×10^6. Levelized production price analysis, also shown in Table XI supports the NPV analysis. These data demonstrate that autocausticizing with borates can be an attractive incremental improvement in chemical pulping. It can compete on a RADR basis. At maturity it can provide a corporation significant benefits as measured by NPV analysis.

The real benefit of autocausticizing is shown best by incremental analysis rather than total system analysis. The reason why incremental analysis is the more accurate barometer is the relative proportion of the savings achieved by autocausticizing compared to conventional kraft pulping. Given the need for incremental analysis, pro forma statements can be constructed for NPV analysis, as shown in Table XII.

TABLE IX. Relative Energy Flows in Borate and Conventional Kraft Mills, Favorable Conditions

Energy requirement (Basis: 1 ton pulp)	Conventional system	Borate system	Difference
Total steam flow	14,856 lb	14,856 lb	–
Btu in steam from wood and liquor	16.1×10^6 Btu	13.4×10^6 Btu	-2.7×10^6 Btu
Btu in oil to lime kiln	2.0×10^6 Btu	–	-2.0×10^6 Btu
Btu in coal to power boiler[a]	3.1×10^6 Btu	6.3×10^6 Btu	3.2×10^6 Btu
Total, all fuel	21.2×10^6 Btu	19.7×10^6 Btu	-1.5×10^6 Btu
Total, fossil fuel	5.1×10^6 Btu	6.3×10^6 Btu	1.2×10^6 Btu

[a]*Due to potential for blackouts, this may have to be treated as oil to the liquor recovery boiler, as shown in the sensitivity analysis.*

Source: Aho [3].

TABLE X. Comparison of Costs, Unbleached Conventional Kraft and Borate Pulping Systems (Basis: 1000 Ton/Day Mill, 1984 $ $\times 10^3$)

	Mill type		
Cost	Conventional kraft	Borate	Difference
Capital costs			
Directs	338,500	319,190	(19,310)[a]
Indirects @ 30%	101,500	95,760	(5,740)
Total	440,000	414,950	(25,050)
Operating costs			
Wood	39,050	39,050	–
Labor	20,940	20,676	(264)
Chemicals	6,230	7,150	920
Oil @ $5/Btu $\times 10^6$	3,750[b]	–	(3,750)
Coal @ $2/Btu $\times 10^6$	2,170	4,380	2,210
Electricity @ 6¢/kWh	2,860	2,860	–
Maintenance @ 3% of capital cost	13,200	12,450	(750)
Supervision and overhead	22,490	22,490	–
Total	110,690	109,056	(1,634)
Cost per annual ton			
Capital	1,257	1,186	71
Operating	316	312	4

[a]*Values in ( ) are treated as negative numbers.*

[b]*This author assumes that no added oil will be required in the black liquor boiler for stability, and that coal can be added in the power boiler. This was tested in sensitivity analysis.*

Sources: Adapted from Aho [1]; Aho [3].

TABLE XI. Financial Analysis of Autocausticizing Based Kraft Pulp Mill[a] (Basis: 1000 Ton/Day Mill, 1984$)

Parameter	Value
Nominal discount rate[b]	
Risk adjusted	*19%*
Mature	*15%*
Nominal levelized production price	
Risk adjusted	*$698/ton*
Mature	*$681/ton*
[Conventional kraft]	*[$696/ton]*
Nominal Net Present Value[c]	
Risk adjusted discount rate	*($1,090,000)[d]*
Mature discount rate	*$17,480,000*

[a]*Total mill analysis rather than incremental analysis unbleached kraft mill.*

[b]*For conversion to real dollar analysis assume inflation is 5%/yr.*

[c]*Based upon levelized production price for conventional kraft.*

[d]*Values in ( ) are treated as negative numbers.*

These pro forma statements employ a levelized cost of conventional kraft pulp of $700/ton as the basis for revenue estimation. Further, they recognize that depreciation becomes a negative cash flow, because a smaller investment provides for a reduction in tax avoidance capability. Year 0 economic values, therefore, also are shown in Table XII.

The use of incremental analysis necessarily implies that the pioneer plant discount rates must be used in the evaluation. Such discount rates are used to measure the benefits of the innovation to the first firm installing autocausticizing. The NPV also has been calculated using mature industry discount rates. This latter calculation demonstrates the economic penalty imposed upon firms not employing autocausticizing, assuming maturation of the technology. The incremental analysis results are shown below. Representative years on the pro forma statement are shown in Table XIII. These cash flows are displayed complete in Fig. 9.

*TABLE XII. Calculation of Incremental Values for Use in
Evaluating Autocausticizing with Borates
(Basis: 1000 Ton/Day Mill)*

Parameter	Conventional kraft mill	Autocausticizing mill	Incremental value
Capital cost	440×10^6	415×10^6	25×10^6
Revenue[a]	245×10^6	245×10^6	–
O & M	110.6×10^6	109.2×10^6	1.4×10^6

[a] *$700/ton as levelized $\times$ 350 $\times 10^3$ tons.*

The cash flows shown in Table XIII and Fig. 9 illustrate
some of the unique characteristics of incremental analysis.
The most critical issue in incremental analysis is that capital investments may become positive cash flows as a result of
capital investment savings. The autocausticizing system involves a capital savings (or positive cash flow) of $26 mil-
lion. When cash flows from investments are positive, cash
flows from depreciation and consequent tax avoidance may be
significantly negative. Capital savings have consequential
reductions in tax avoidance.

The NPVs associated with borate based autocausticizing
have been calculated over a range of discount rates, and
these are shown in Fig. 10. These NPVs demonstrate that
autocausticizing is promising regardless of discount rate.[1]
Mathematically, if the analysis employs the discount rate
range of 25-150, the NPV reaches its nadir. Using lower rates
of discount results in higher NPV results. Very high rates
of discount also result in high NPVs. As the discount rate
approaches infinity, the NPV approaches the value of the
negative capital investment, or capital savings.

The only conclusion to be drawn from the data presented
above is that autocausticizing is highly attractive. It can
generate both capital and operating cost savings. As the

---

[1]*Coincidentally, this analysis highlights one of the
problems with rate of return type analysis. When incremental
evaluations are made, it is common to have negative investment (capital savings) values, or positive cash flows from
investments. If there are substantial operating year positive cash flows as well, no solution to the internal rate of
return calculation is mathematically possible. Under such
circumstances NPV is the only accurate measure of the worth
of an investment.*

TABLE XIII. Cash Flows for Autocausticizing in Selected Years

Project Year	0	1	2	3
Calendar Year	1984	1985	1986	1987
Revenue		0	0	0
O & M (fixed)		1,065	1,118	1,174
O & M (variable)		650	683	717
Depreciation		5,010	8,016	6,012
Earnings before taxes		6,725	9,817	7,903
Taxes		3,094	4,516	3,635
Net income after taxes		3,632	5,301	4,268
Depreciation		(5,010)[a]	(8,016)	(6,012)
Investment[b]	(25,050)			
Investment tax credit[b]	(2,505)			
Nominal net after tax cash flow[b]	22,545	(1,378)	(2,715)	(1,744)

[a] *Values in ( ) are treated as negative numbers. Values for operating and maintenance costs above the tax line are operational savings that generate positive tax requirements and cash flows.*

[b] *Investments not made, or capital savings.*

incremental analysis shows, there is no circumstance when autocausticizing is not attractive. On a NPV basis, it always yields a positive result.

The analysis of autocausticizing as presented above assumes that the reduced heat content of the spent pulping liquor from the borate based system does not cause flame stability and potential blackout problems in the liquor recovery boiler. Consequently the analysis presented above assumes that the steam not raised by the liquor recovery boiler can be generated by burning coal in the power boiler. There is serious concern that the liquor recovery boiler will have flame stability and blackout problems in a borate based autocausticizing system [3]. Because of such concerns, the

4	5	6	10	20	30
1988	*1989*	*1990*	*1994*	*2004*	*2014*
0	*0*	*0*	*0*	*0*	*0*
1,233	*1,294*	*1,359*	*1,652*	*2,690*	*4,382*
753	*791*	*830*	*1,009*	*1,644*	*2,677*
4,008	*2,004*	*0*	*0*	*0*	*0*
5,994	*4,089*	*2,189*	*2,661*	*4,334*	*7,060*
2,757	*1,881*	*1,007*	*1,224*	*1,994*	*3,248*
3,237	*2,208*	*1,182*	*1,437*	*2,340*	*3,812*
(4,008)	*(2,004)*	*0*	*0*	*0*	*0*
(771)	*204*	*1,182*	*1,437*	*2,340*	*3,812*

flame stability assumption has been subjected to economic
sensitivity analysis.

In order to test the impact of flame stability and blackout problems on the economics of autocausticizing, it was
assumed that the power boiler could not raise the steam "lost"
due to a less efficient recovery boiler. It was assumed that
such steam would have to be raised by burning oil, costing $5/
5/Btu $\times$ 10^6, in the recovery boiler. When this more con-
servative fuel substitution assumption is made, the use of
autocausticizing results in a net fuel cost increase of
$1,775,000 with respect to conventional kraft mills. How-
ever the 1000 ton/day mill experiences no change in wood
costs; an increase in chemical purchases of $920,000; and a
decrease in the costs of labor, maintenance, and supervision

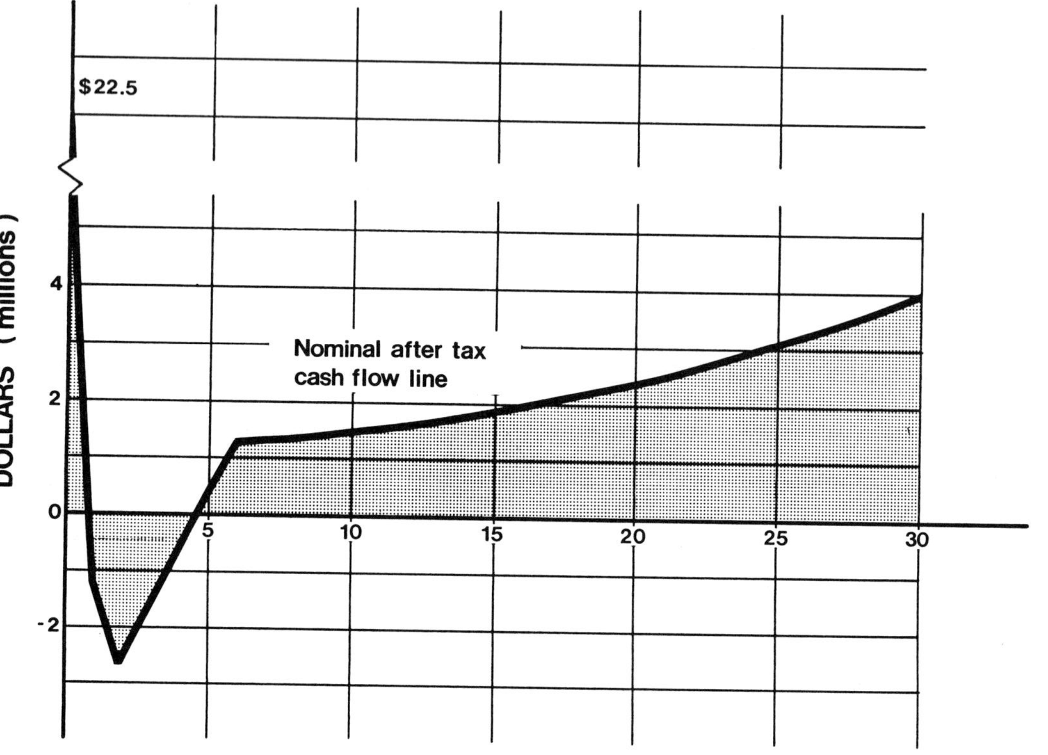

FIGURE 9.   The cash flows from the pro forma statement for autocausticizing.

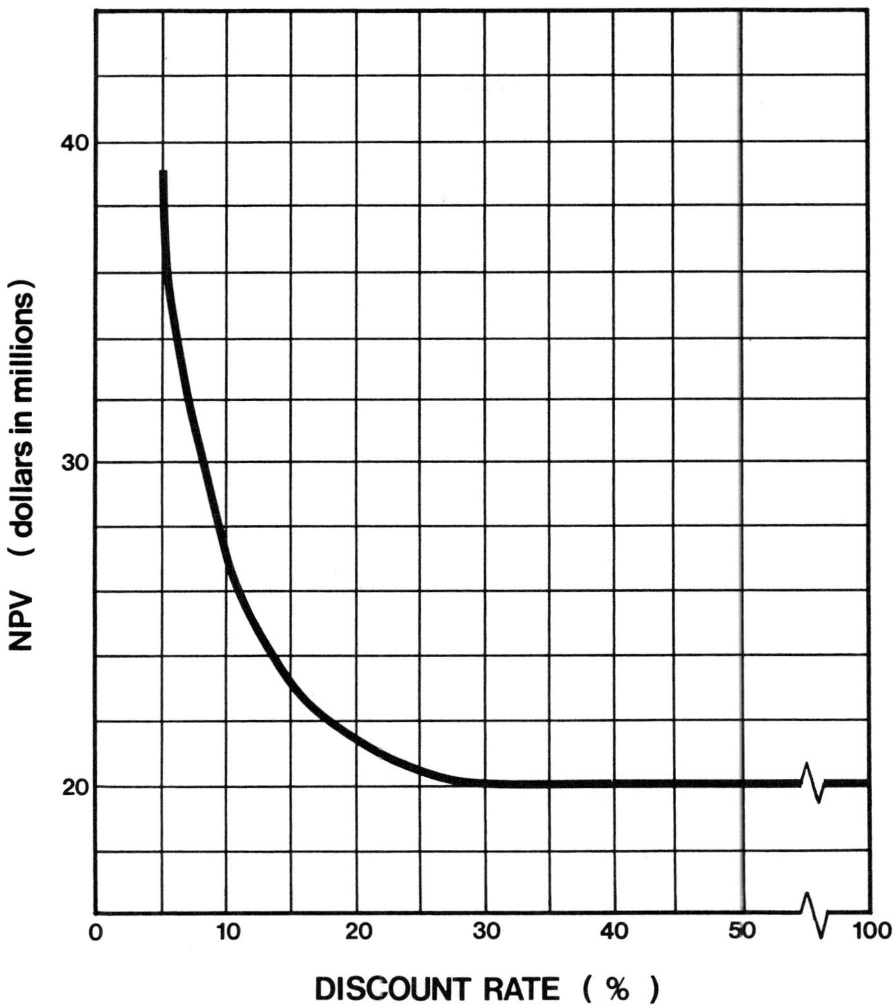

FIGURE 10. Net Present Values for autocausticizing
as a function of the discount rate.

totalling $2,648,000 when using borate based autocausticizing.
Therefore operating costs at the mill increase by $47,000/yr
(net) or $0.13 ton of pulp under the most conservative
assumptions.

The $47,000/yr or $0.13/ton increase in annual operating
costs as calculated above represents virtually no difference
in annual operating costs between autocausticizing and conventional kraft pulping. Because there is virtually no difference in annual operating costs, the NPV of

autocausticizing with borates becomes a function of the influence of inflation and the discount rate on the value of
the tax avoidance impacts of depreciation. Under any condition where inflation is greater than zero, or where the discount rate is greater than zero, the NPV of autocausticizing
will remain positive for greenfield mill applications. The
consequence of the absence of any difference in annual costs
between autocausticizing with borates and conventional kraft
pulping, also, is an absence of incentive to convert existing
kraft mills to autocausticizing if flame stability problems
do occur in the liquor recovery boiler.

A probability of 0.5 may exist for either fossil fuel
autocausticizing scenario, depending upon current mill trials.
Autocausticizing remains as a high desirable improvement for
the kraft process, at least for new mills, regardless of
which assumption is made concerning makeup fuel. Autocausticizing with borates is attractive because it achieves process
simplification and consequent capital investment savings.
It eliminates the need for a lime kiln and the associated
recausticizing equipment associated with the conventional
kraft pulp mill.

B. Pulping with Organic Solvents (Organisolv Pulping)

Organisolv pulping is a more radical departure from conventional chemical pulping than autocausticizing. A sulfur
free alcohol based pulping liquor is substituted for the
$NaOH + Na_2S$ based kraft liquor.

1. Pulping Chemistry. Pulping of both hardwoods and
softwoods in alcohol has been studied for some time. Patents
covering such processes date back to 1932 [37]. Today some
nine patents have been issued concerning this concept. In
principle, wood chips are subjected to a liquor containing
20 to 80% alcohol (wt. basis) at temperatures of 160–220°C
(320–430°F) and pressures of 10 to 50 atm. (150–750 psi) in
order to achieve delignification [38]. The alcohol may be
MeOH, EtOH, or butanol. Catalysts such as aluminum salts may
be used to facilitate the process [31].

Preferred conditions of organisolv pulping appear to be
360–410°F (180–210°C) and 300–525 psi (20 to 35 atm.) pressure. The preferred pulping liquor contains MeOH or EtOH at
concentrations of 40–60% (wt. basis) and water [38]. Mech-
anistically, under these conditions, organisolv delignification may proceed by attacking β-0-4 linkages in the lignin
and the lignin-hemicellulose bonds [31]. The predominant
β-0-4 linkages do not appear to be readily attached in
organisolv systems.

2. Process Considerations. Organisolv pulping offers yields in the region of 50 to 55% [22, 23, 33], consistent with the best yields of hardwood kraft or sulfite pulping. At the same time pulping in organic solvents produces pulps with strength properties more closely approximating sulfite pulps than softwood kraft pulps [22, 31] although strength properties approaching those associated with hardwood kraft have been reported on one occasion [23]. On that basis, organisolv pulps are comparable in quality to the products of the CTMP processes.

The distinguishing feature of the organisolv process, then, is its ability to be energy self-sufficient as is shown in Fig. 11. The spent liquor from organisolv may be processed to recover the lignin as a powder, containing 11,500 Btu/lb. The hemicelluloses may be extracted as a sugar solution. Hemicelluloses may be processed by anaerobic fermentation to produce methane gas. Some of the methane produced can fuel gas fired internal combustion engines used for power generation and feedwater heating and these can supplement conventional lignin fired steam boiler cogeneration systems. The surplus methane can be sold as SNG [33]. Alternatively the hemicelluloses and lignin can be used as feedstocks to produce chemicals.

3. Organisolv Economics. Organisolv pulping has attracted considerable attention as a way to build small (e.g., 150 ton/day) increments of capacity [23] as a way to effect capital savings, and as a way to effect operating cost savings through the sale of energy rather than the purchase of fossil fuels and electricity. As a consequence, it has been proposed for commercialization by such firms as Biological Energy Corp. and Foster Wheeler Energy Corp. [4]; and by General Electric Venture Capital Corp. as an investor in Biological Energy Corp. [13].

Much of the economic advantage attributed to organisolv pulping stems from its apparent low capital cost. This lower capital cost is the basis for some of the current excitement concerning organisolv pulping. There is considerable disagreement about those capital costs, however, between Katzen, Frederickson, and Brush [22], General Electric [13], and Paszner and Chang [29]. The Paszner values, for example, deliberately double the Katzen values. As a consequence of this disagreement, an order-of-magnitude capital cost estimate has been generated here for a 1000 ton/day ethanol based organisolv mill, and it is shown in Table XIV. It is substantially higher than either the Katzen or Paszner estimate, however it is substantially lower than the cost of a kraft mill, and the capital cost of a kraft mill is virtually identical to the capital cost of a sulfite mill [11]. This

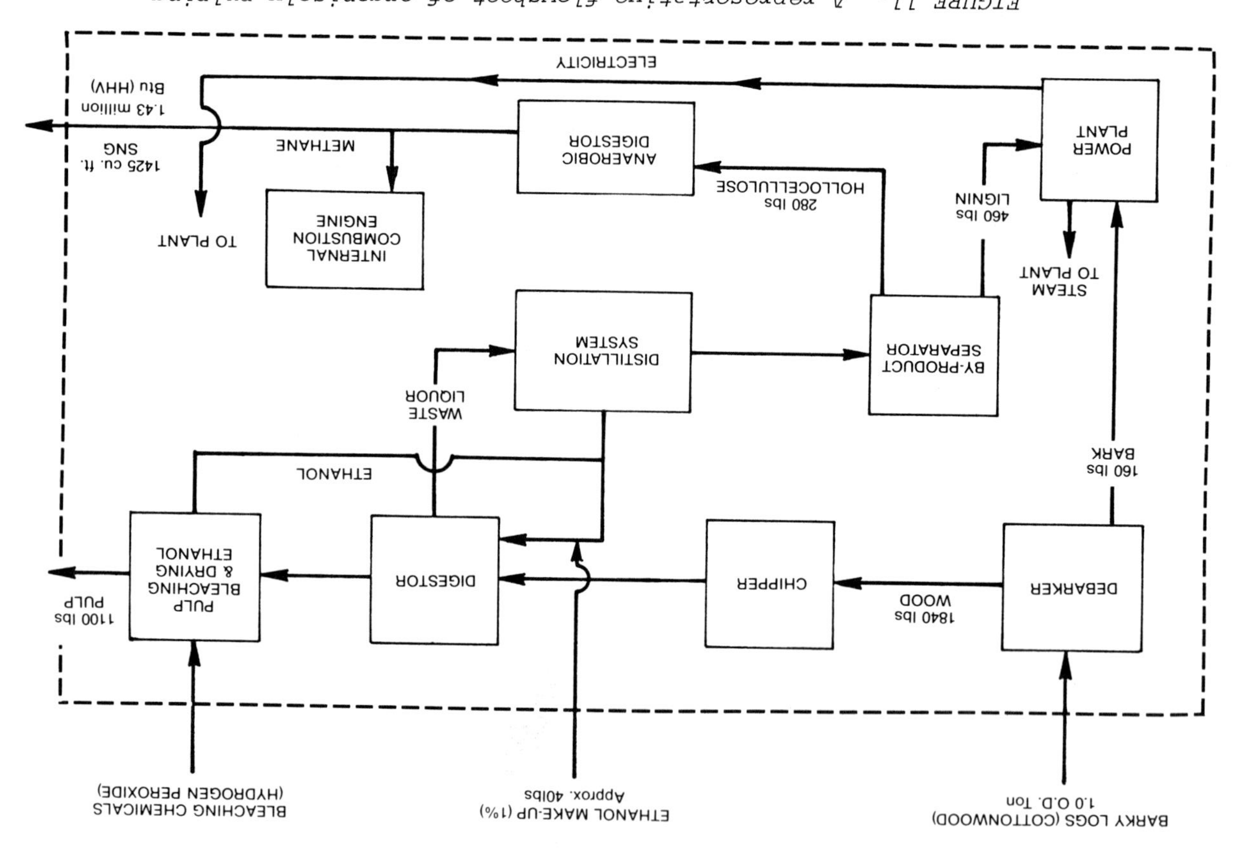

FIGURE 11. A representative flowsheet of organisolv pulping identifying major fiber and energy flows. Source: [33].

*TABLE XIV.   Comparative Capital Cost Estimate, 1000 Ton/
Day Kraft and Organisolv Mills (1984 $ × 10³)*

Cost Category	Kraft	Mill organisolv	Difference
Direct costs			
General facilities	23,700	23,700	0
Wood handling	40,620	40,620	0
Pulping, washing and screening	50,780	50,780	0
Pulp drying	47,390	47,390	0
Black liquor separation and digester	0	15,650	15,650
Liquor recovery boiler	81,240	0	(81,240)[a]
Power generation			
Solid fuels boiler and turbine	47,380	60,790	13,410
Diesel generator		2,710	2,710
Recausticizing system	20,310	0	(20,310)
Wastewater treatment	27,080	13,540	(13,540)
Subtotal	338,500	255,180	(83,320)
Indirect costs			
Engineering @ 10%	33,850	25,520	(8,330)
Construction management @ 10%	33,850	25,520	(8,330)
Working capital @ 5%	16,900	12,800	(4,100)
Contingency @ 5%	16,900	12,800	(4,100)
Total	440,000	331,820	(108,180)

[a]*Values in ( ) are treated as negative numbers.*

TABLE XV.   Cost Comparison of Organisolv and Related
Pulping Systems (Basis: 1000 Ton/Day Mill, 1984$)

		Cost	Cost
Pulp mill type	*Capital ($\times 10^3$)*	*($/annual ton)*	*Yield (%)*
Organisolv			
Author estimate	*331,820*	*950*	*55[c]*
Katzen estimate[b]	*32,130*	*610*	*55[c]*
Sulfite	*430,800*	*1,230*	*55*
Chemithermomechanical pulping	*215,000*	*610*	*85*
Hardwood kraft	*440,000*	*1,260*	*55*

[a]*Not including capital costs.*

[b]*150 t/d mill, escalation factor from M & S equipment
cost index = 1.29.   Furnish not escalated.*

[c]*Based on Katzen et al. [23].*

capital cost estimate is used as an upper bound in the
analysis while the Katzen estimate is considered to be the
lower bound.

Capital, wood, operating, energy, and maintenance costs
have been estimated for organisolv pulping and these values
are presented in Table XV. Similar values for sulfite,
CTMP, and hardwood kraft pulping also are presented in Table
XV. With respect to Table XV, it is useful to note that the
capital costs of organisolv as estimated by Katzen are identical to those associated with CTMP on a $/ton basis. Annual
costs are considerably cheaper, largely due to differences in
energy cost. Organisolv appears to have some advantage over
sulfite and hardwood kraft processes when sulfite byproduct
chemicals are ignored. However, organisolv pulping appears
to have no advantage per se over CTMP processes in producing
conventional pulps.

The engineering economy analysis of organisolv pulping is
presented in Tables XVI and XVII, using pioneer plant discount rates for organic solvent systems, second generation
RADR rates for CTMP, and conventional rates for sulfite and
hardwood kraft pulping. On the basis of these data, it

parameter				

Wood cost ($/ton)	Operating cost ($/ton)	Energy cost ($/ton)	Maintenance cost ($/ton)	Total annual cost[a] ($/ton)
91	125	(13)[d]	31	234
91	68	6	18	183
91	194	30	37	352
59	81	84[e]	18	242
91	162	25	38	316

[d]*Byproduct SNG credit @ $5/Btu $\times 10^6$*

[e]*1400 kWh/ton @ 6¢/kWh*

Sources: Katzen et al. [23]; Paszner and Chang [29]; Aho [1]; Tillman, Rossi, and Simmons [33].

appears that the CTMP process, currently being installed as
a second generation for TMP pulping, will be more attractive
than organisolv within the constraints of conventional pulp
markets. Organisolv may be more suited to the production of
chemicals or foodstuffs from wood with conventional pulp
being, at most, a byproduct.

V. CONCLUSIONS

Significant innovation has occurred within the past two
decades, particularly in the area of mechanical pulping. The
TMP and RMP processes have permitted mechanical pulping of
sawmill residues. At the same time TMP has given mechanical
pulping a higher strength, more versatile product.
Despite the gains made in mechanical pulping, chemical
pulping remains dominant, and the kraft process remains the
most significant pulping system available and chemical pulping is in a serious cost/productivity squeeze.

TABLE XVI. Financial Analysis of Organisolv Pulping

Parameter	Value
Real discount rate	
Risk adjusted	*19%*
Mature	*10%*
Nominal discount rate	
Risk adjusted	*25%*
Mature	*15%*
Real levelized cost, Author estimates	
Risk adjusted	*466*
Mature	*340*
Real levelized cost, Katzen based estimates	
Risk adjusted	*321*
Mature	*245*
Comparative real levelized costs[a]	
Sulfite	*475*
Hardwood kraft	*445*
CTMP	*324*
Nominal levelized cost, Author estimates	
Risk adjusted	*613*
Mature	*532*
Nominalized levelized cost, Katzen based estimate	
Risk adjusted	*422*
Mature	*383*
Comparative nominal levelized costs[b]	
Sulfite	*743*
Hardwood kraft	*696*
CTMP	*471*

[a]*Discount rate = 10%*

[b]*Discount rate = 15%*

TABLE XVII. Comparison of Pulping Technologies

Parameter	Organisolv (Author estimates)	Organisolv (Katzen estimates)[a]	Sulfite[b]	CTMP[b]	Hardwood kraft[b]
Unit size (tons/day)	1,000	150	1,000	1,000	1,000
Appropriate nominal discount rate (%)	25[c]	25[c]	15	16[d]	15
Nominal levelized production price ($/ton)	613	422	743	471	696
Real levelized production price ($/ton)	466	321	475	324	445

[a]Based on [22], [23].

[b]Based on [1], [11].

[c]Pioneer plant discount rate.

[d]Second generation technology discount rate.

CTMP processes are bringing chemical and mechanical
processes closer together in terms of product quality while
retaining the high yields and low capital investment factors
associated with mechanical pulping systems. Other potential
advances in chemical pulping include oxygen delignification,
the various autocausticizing systems (e.g., borates, DARS),
organisolv pulping, amine pulping, and ultimately biological
pulping.

Two processes have been investigated in some detail:
(1) borate based pulping, and (2) organisolv pulping. Inferentially CTMP also was evaluated as it is a competitor with
sulfite pulping today. While the autocausticizing processes
appear to have significant potential to improve pulp production, the organisolv processes appear less attractive for
the manufacture of pulp. Researchers must seek a quantum
breakthrough in pulping economics in order to displace kraft
pulping and inject growth industry vigor into pulping and
bleaching.

REFERENCES

1. Aho, W. O. 1983. Advances in Chemical Pulping
   Processes. *In* "Progress in Biomass Conversion," Vol. 4.
   Academic Press, New York.

2. Aho, W. O. 1984. Recent Developments in Chemical Pulping Technology. Presented at the 1984 Annual Meeting of
   the American Institute of Chemical Engineers, San
   Francisco, Nov. 25-30.

3. Aho, W. O. 1984. Personal communication, Nov. 14.

4. Anon. 1983. Solvent Pulping to be Commercialized.
   Chemical and Engineering News 61(49):38.

5. Babcock and Wilcox. 1978. "Steam: Its Generation and
   Use," 39th Ed. Babcock and Wilcox, New York.

6. Basta, N., Kummant, I., Schabas, W., Froelich, G., Skole,
   R., and Tikkanen, T. 1984. New Pulping Methods—On
   Paper They Look Great. *Chemical Engineering 91*(10):22-26.
   Also see Cody, H. M. 1984. Quesnel River Pulp Converts
   TMP Mill to a 450 tpd CTMP Operation. *Pulp and Paper
   58*(6):76-77.

7. Bethel, J. S. *et al.* 1976. Renewable Resources for
   Industrial Materials. National Research Council, National Academy of Sciences, Washington, D.C.

8. Brink, D. L., Charley, J. A., Faltico, G. W., and
    Thomas, J. F. 1976. The Pyrolysis-Gasification Combus-
    tion Process: Energy Considerations and Overall Process-
    ing. *In* "Thermal Uses and Properties of Carbohydrates
    and Lignins." Academic Press, New York.

9. Brink, D. L., Faltico, G. W., and Thomas, J. F. 1977.
    The Pyrolysis-Gasification-Combustion Process: Energy
    Effectiveness Using Oxygen vs Air with Wood Fueled
    Systems. *In* "Fuels and Energy From Renewable Resources."
    Academic Press, New York.

10. Coleman, M. 1984. Lime Kiln Converted to Burn 50% Com-
    bustible Gas From Woodwaste. *Pulp and Paper* 58(11):140-
    143.

11. Ekono, Inc. 1982. Personal communication from W. Aho
    and K. Leppa, Mar. 15. Also see: Kojo, M. 1979.
    Profitability of Recausticizing. Paperi ja Puu — Papper
    och Tra 61(9):581-583.

12. Frederick, W. J. 1984. The Energy Costs of Increased
    Organics Recovery for Chemical Byproducts in Kraft Pulp
    Mills. *In* "Progress in Biomass Conversion," Vol. 5.
    Academic Press, New York.

13. General Electric. 1983. The Opportunity: Tree Refining.
    Biological Energy Corporation/GEVENCO.

14. Goel, K., Ayroud, A. M., and Branch, B. 1980. Anthra-
    quinone in Kraft Pulping. TAPPI 63(8):83-85.

15. Grace, T. M. 1979. Chemical Recovery Technology. 32nd
    Annual Seminar of the Pacific Section of TAPPI, Seattle,
    Washington, Sept. 27-28.

16. Granskog, J. 1978. Economies of Scale and Trends in
    the Size of Southern Forest Industries. *In* "Complete
    Tree Utilization of Southern Pine," Proc. C. McMillin,
    ed.). Forest Products Research Society, Madison,
    Wisconsin.

17. Hersh, H. N. 1981. Energy and Materials Flows in the
    Production of Pulp and Paper. Argonne National Labora-
    tory, Chicago, Illinois.

18. Hurley, P. J. 1978. Comparison of Mill Energy Balance
    Effects of Conventional, Hydropyrolysis, and Dry Pyroly-
    sis Recovery Systems. Institute of Paper Chemistry,
    Appleton, Wisconsin.

19. Janson, J. 1977-1979. The Use of Unconventional Alkali
    in Cooking and Bleaching. Part 1. A New Approach to
    Liquor Generation and Alkalinity. Part 2. Alkali Cook-
    ing of Wood with the Use of Borate. Part 3. Oxygen-
    alkali Cooking and Bleaching with the Use of Borate.
    Part 4. Kraft Cooking wth the Use of Borate. Part 5.
    Autocausticizing Reactions. Part 6. Autocausticizing of
    Sulphur-Containing Model Mixtures and Spent Liquors.
    Paperi ja Puu — Papper och Tra. 59(7):425-429; 59(9):
    548-557; 60(2):97-100; 60(5);349-352, 355-357; 61(1):20-
    30; 61(2):98-103.

20. Jansen, J. 1979. Autocausticizing Alkali and Its Use in
    Pulping and Bleaching. *Paperi ja Puu — Papper och Tra.*
    61(8):495-504.

21. Jansen, J. 1980. Pulping Processes Based on Autocausti-
    cizable Borate. *Svensk Papperstidning 14*:392-395. Also
    see Jansen, J. 1980. Pulping Processes Based on Auto-
    causticizable Borate. Eucepa Symposium. The Delignifi-
    cation Methods of the Future. Helsinki, Finland.

22. Katzen, R., Frederickson, R., and Brush, B. 1980. The
    Alcohol Pulping and Recovery Process. *Chemical Engineer-
    ing Progress*, Feb., 62-67.

23. Katzen, R., Frederickson, R., and Brush, B. 1980.
    Alcohol Pulping Appears Feasible. *Pulp and Paper 54*(8):
    144-149.

24. Law, K. N., Rioux, P., LaPointe, M., and Valade, J. L.
    1984. Chemi-thermomechanical Pulping of White Birch for
    Newsprint. Forest Products Research Society Annual
    Meeting, St. Louis, June 24-28.

25. Leppa, K. 1980. "Swedish Pulp Mill Switches From Oil
    to Woodwaste and Peat." Energy Systems Guidebook.
    McGraw-Hill, New York.

26. McDonough, T. J., and Van Drunen, V. J. 1980. Pulping
    to Low Residual Lignin Contents in the Kraft-
    Anthraquinone and Kraft Processes. TAPPI 63(11):83-87.

27. Metsavirta, A., and Leppanen, M. 1980. Improved Pro-
    fitability of Thermomechanical Pulping Using Computer
    Control and Effective Heat Recovery. TAPPI 63(7):37-41.

28. Minor, J. 1982. Pulp. *In* Kirk-Othmer "Encyclopedia of
    Chemical Technology," 3rd Ed., Vol. 19. John Wiley and
    Sons, New York.

29. Paszner, L., and Chang, R. Undated. Biomass Refinery: Catalyzed Organisolv Process for Total Chemical Conversion of Wood Into High-Yield Pulp and Chemical Byproducts.

30. Sanyer, N. 1982. Status of New Pulping Processes: Problems and Prospects. Forest Products Laboratory, U.S. Department of Agriculture Forest Service, Madison, Wisconsin.

31. Sarkanen, K. V. 1980. Acid-Catalyzed Delignification of Lignocellulosics in Organic Solvents. *In* "Progress in Biomass Conversion," Vol. 2. Academic Press, New York.

32. Sjostrom, E. 1981. "Wood Chemistry: Fundamentals and Applications." Academic Press, New York.

33. Tillman, D., Rossi, A., and Simmons, S. 1982. Wood: Its Present and Potential Uses. Envirosphere Co., Bellevue, Washington. For the U.S. Congress, Office of Technology Assessment.

34. Tillman, D. A. 1983. Evaluating the Potential Contribution of the Forest Products Industry to U.S. Energy Supply. *Journal of Applied Polymer Science: Applied Polymer Symposium 37*:593–615.

35. Tillman, D. A. 1983. Making the Best Energy Use of Wood. *In* "Progress in Biomass Conversion," Vol. 4. Academic Press, New York. Also see Tillman, D. A. 1983. Maximizing Efficiency of Wood for Product and Energy. *In* "Energy Technology X," Proc. Government Institutes, Inc., Washington, D.C.

36. U.S.D.A. Forest Service. 1981. U.S. Timber Production, Trade, Consumption, and Price Statistics: 1950–1980. U.S. Government Printing Office, Washington, D.C.

37. U.S. Patent #1,856,567.

38. U.S. Patent #4,100,016.

39. Walker, P. J., and Batsis, E. J. 1978. Heat Recovery From TMP Operations Results in Energy Conservation. *Pulp and Paper 52*(3):146–148.

40. Westergaard, B. 1984. B. C. Mill Successfully Fires Lime Kiln With Wood Waste. *Pulp and Paper Journal 37*(6): 20–22.

41. Westling, J. E., and McKean, W. T. 1982. The Direct Alkaline Recovery System (DARS), An Alternative to Kraft Recovery. Presented at 1982 Annual Meeting, American Institute of Chemical Engineers, Los Angeles, Nov. 14–19.

Chapter VI

ADVANCED PROCESSES FOR LUMBER MANUFACTURING

I. INTRODUCTION

While sawmilling is no longer the most dominant forest
industry, it is still prominent. Sawmilling was the first
forest industry manufacturing process to emerge. Even today
lumber manufacturing is second only to pulp production as a
product sector of the FPI. Three of the four largest forest
industry companies began as lumber and board producers:
Champion-St. Regis (if merged), the Weyerhaeuser Company,
and the Georgia-Pacific Company. Consequently advanced
processes for the production of lumber and board products are
of substantial consequence to the industry.

Lumber manufacturing is used here, inferentially, to
represent the production of all board and panel products
(including plywood). Such a representation does not totally
ignore tremendous advances in the manufacture of structural
composite panel products including particleboard cored ply-
woods (or veneer skinned particleboards), flakeboards, and
other rapidly advancing products. The production of such
engineered composite panel products is approaching certain
exciting breakthroughs including the use of lignin and spent
pulping liquor based binders (see, for example, Soltes and
Lin [35]) and ultimately the use of chemically activated lignin within the wood chips as a binder [7, 22]. However such
composites typically compete with other wood products such as
plywood, rather than nonwood materials such as metals or
plastics. Further, such products are generically similar to
lumber alternatives.

The area of lumber manufacturing is viewed broadly here
to include all products serving the same market as conventional clear lumber. As stated previously, this portion of
the industry faces traumatic problems in switching from old
growth technology, as depicted in Fig. 1, to systems for
processing of small logs. Small logs, shown being harvested
in Fig. 2, present unique problems to the industry. These
problems are in addition to the overcapacity of the industry,

FIGURE 1. *A large log, old growth sawmill in the Pacific Northwest, symbolizing technological obsolescence.*

FIGURE 2. *Harvesting small logs in the Southeast.*

and the projected stagnation of the housing market. There
are a number of new processes in this area designed to overcome the cost/price pressures induced by the overcapacity and
nonwood competition problems discussed in Chapter I. Such
processes are surveyed here, and then one is selected for
detailed analysis using the principles and data developed in
Chapters II-IV.

II. THE STATE-OF-THE-ART IN LUMBER MANUFACTURING

In order to evaluate advanced processes in the manufacture of lumber products, it is necessary to examine the current production systems. Such an examination includes principles, processes, costs, and economic pressures associated
with the current sawmill.

A. *Principles of the State-of-the-Art Sawmill*

The state-of-the-art sawmill must be viewed as a multiproduct plant. Its objective is not to maximize lumber production. Its objective is to maximize cash flow generation
through the production and sale of lumber, pulp chips, particleboard furnish, fuel, and specialty products (e.g.,
beauty bark, sawdust for cattle food, chicken litter, etc.).
Ideally the sawmill both maximizes cash flows and, to a great
extent, stabilize its cash flows by serving countercyclical
markets with secondary products. Within the context of such
cash flow management, it is useful to keep certain approximate product value relationships in mind. Simplified relationships are presented in Table I. Particleboard furnish
prices are in between the fuel and pulp chip values shown in
Table I. Such values aid in determining the interplay of
yields by sawmill product.
This concept of the sawmill as a multiproduct cash flow
generator is most important. When it emerged (see Chapter
IV) it provided the basis for the establishment of integrated sawmill-pulp mill complexes. Also it provided strong
support for many innovations (e.g., particleboard, TMP pulping) as explored in Chapter IV.
Within this multiproduct concept, it is important that
the conventional sawmill changes only the physical shape and
perhaps the moisture content of its final products. It does
not change the chemical composition, organization, or internal bonding of the fibers within the wood produced in such
forms as lumber, pulp chips, or fuel.

TABLE I. Typical Relative Values for Sawmill Products

Product	Typical value (1984 \$/O.D. ton)
Lumber	$230 - 290^a$
Pulp chips	$45 - 55^b$
Fuel	$20 - 35$

$$^a \quad \frac{58 \ ft^3}{MBF} \times \frac{30 \ lb, O.D.}{ft^3 \ O.D.} \times \frac{ton}{2000 \ lb} = \frac{.87 \ ton, O.D.}{MBF}$$

then,

$$\frac{\$200-250}{MBF} \quad \frac{MBF}{0.87 \ ton} = 230-290$$

^{b} \$40-50/cord. This can be as high as \$100/O.D. ton.
Sources for base prices: [28, 40].

B. *State-of-the-Art Sawmilling Processes*

The small log sawmill, as described completely by Williston [42], and by other authors (see for example Haygreen and
Bowyer [17]), represents the state-of-the-art in lumber processing today. This mill is reviewed in cursory fashion
below.

1. *Unit Operations of the State-of-the-Art Sawmill.* The
small log sawmill includes the following unit operations or
machine centers: (1) debarking, (2) log merchandising, (3)
primary breakdown (the headrig), (4) board edging, (5) board
trimming, (6) product drying, (7) product planing, (8) pulp
chip production, and (9) fuel production (hogging). Although
mill controls are not a unit operation per se, they are a
necessary consideration within the individual component context. The unit operations where major advances have occurred
in the past 10-15 years include: (1) log scanning and merchandising, (2) the headrig, (3) edging, (4) drying, (5) planing, and (6) system controls.
In the most advanced small log common lumber mills, logs
are yarded, debarked, run through the cutoff saw shown in
Fig. 3, and sorted by log dimension during the merchandising
phase of activity. Logs are debarked in ring or rosserhead
systems. Ring debarkers tend to offer higher yields (lower
white wood losses), and therefore they are preferred [38].
Sorting as shown in Fig. 4 is performed with respect to

FIGURE 3. The cutoff saw at a small log sawmill. (Photo courtesy of W. R. Smith.)

FIGURE 4. Log sorting for mill efficiency. (Photo courtesy of W. R. Smith.)

species, large end and small end diameters, taper, sweep, and
other relevant characteristics. Merchandising may be performed with the aid of scanners for precise shape determina-
tion [41]. The purpose of such sorting activities is to
improve the economic efficiency of the sawmill by facilita-
ting both yield and throughput [36]. Merchandising may proceed at a rate of 3-6 logs per minute [12].

Debarked logs then proceed to the headrig. There are
numerous types of headrigs, and numerous conveying systems
associated with those headrigs. Conveying/holding systems
include alligator infeed systems, sharp chains that impale
small logs, and log chargers feeding carriages [1]. Carriages may be tilted in order to improve recovery by more
secure dogging, using gravity to aid in holding the logs
against the knee [18]. Carriages also may use end dogging
systems in order to improve throughput rates.

Headrig cutting systems include such options as chippers,
bandmills, circular saws, scrag saws, and log gang saws. Of
these chipping headrigs and bandmills, and combinations of
the two, appear most popular. Chipping headrigs are favored
for high throughput rates (e.g., 7-9 logs/min compared to as
low as 3-4 logs/min for bandmills). The lumber recovery of
chipping headrigs with alligator feed systems may be 10% less
than that associated with bandmills, however [1]. Such chipping headrig systems are shown in Figs. 5 and 6. Further,
chipping headrigs are severely penalized by log sweep.
Williston [42] provides a detailed review of this penalty.

The apparent penalty in yield of chipping headrigs as
opposed to bandmills is not as severe as first appears, however. Chipping headrigs tend to produce a higher grade of
lumber. Snellgrove [34], for example, provides a yield by
grade comparison presented in Table II. This yield by grade
advantage may explain why over 25% of the headrigs now in use
are of the chipping headrig type.

Headrigs have undergone a revolution in controls, initi-
ated by the development of the Best Open Face (BOF) computer
cutting program of the U.S. Forest Service, Forest Products
Laboratory, Madison, Wisconsin, that optimizes the yield from
a given log. In the modern mill the BOF program is coupled
to photoelectric or laser sensing devices that profile the
log. The BOF program is at the heart of an integrated headrig consisting of the infeed system, the cutting system, and
the microprocessor. The headrig also may emit laser lines to
aid the sawyer visually as he makes his first cuts. These
controls, depicted in Fig. 7, have increased lumber yields
significantly.

Following the headrig, the wood is processed by subsequent saws, by edgers and by trimmers. Edgers such as the
one depicted in Fig. 8 may be computer controlled systems

FIGURE 5. A chipping headrig. (Photo courtesy of
W. R. Smith.)

FIGURE 6. Infeed to the chipping headrig. (Photo
courtesy of W. R. Smith.)

TABLE II.   Lumber Yields by Grade as a Function of Headrig Type

	Grade		
Headrig type	*Select structural (%)*	*Construction (%)*	*Construction and better (%)*
Beaver[a]	*13.7*	*70.4*	*84.0*
Chip-N-Saw[a]	*34.6*	*46.5*	*81.1*
Bandmill	*6.6*	*63.8*	*70.4*

[a]*Chipping headrig types.*

Source: Snellgrove [34].

FIGURE 7. Headrig control with BOF program. (Photo courtesy of W. R. Smith.)

FIGURE 8. The edger at a small log mill. (Photo
courtesy of W. R. Smith.)

that detect wane in the board or studs by differences in air
pressure. They may remove wane either by chipping cutter
heads or by thin band saws. Like the headrigs, edgers are
controlled by microprocessor or computer. Also like the
headrig, the edgers and trimmers are major producers of pulp
chips, the vital secondary product illustrated by Fig. 9.
    Edge and trim operations typically are followed by the
dry kiln. Both steam and hot flue gas kilns have been used,
with steam being dominant. In order to maximize throughput,
high temperature kilns including continuous rising temperature (CRT) kilns demanding 150 psig steam tend to be preferred over slower drying rate kilns using 15 psig steam.
The high temperature and CRT kilns can reduce drying time by
a factor of two (2.0) to three (3.0).
    The dry kiln is followed by planing operations. While
the conventional cutting planer predominates, abrasive planing has been introduced. Abrasive planing reduces the losses
at this machine center. Abrasive planing produces a fuel
grade residue however, sanderdust. Conventional planing
generates a particleboard furnish grade residue, shavings.
If the dry kiln and planer are eliminated, the mill produces
rough green lumber. The kiln and planer are optional

*Figure 9. Pulp chip storage and transport from the
small log sawmill. (Photo courtesy of W. R. Smith.)*

devices, depending upon market conditions, designed to increase the value of the product manufactured.

   2. Yields of the State-of-the-Art Sawmill. Currently
the average sawmill has a lumber recovery of about 40% on a
ft^3 basis [20],a value which approximates the yield predicted
by the Committee on Renewable Resources for Individual Materials (CORRIM) for the year 1985 [6]. Comparisons of these
yield estimates are shown in Fig. 10. Note that the major
differences appear to be in the distribution of residues between particleboard and fuel.
   In contrast to the yields postulated for "average mill,"
state-of-the-art mills can do considerably better. Assuming
10% bark on the logs, yields can be about 50% (45-54% ft^3
basis) based upon logs up to 15 in. diameter [16]. The primary variable is the headrig and cant breakdown system as
shown in Table III.
   Dealing with averages can be misleading because it masks
such previously mentioned variables as log diameter, sweep,
crook and taper; lumber product and quality produced; and
other conditions. It ignores such variables as species being
processed, markets being served, and other critical issues as

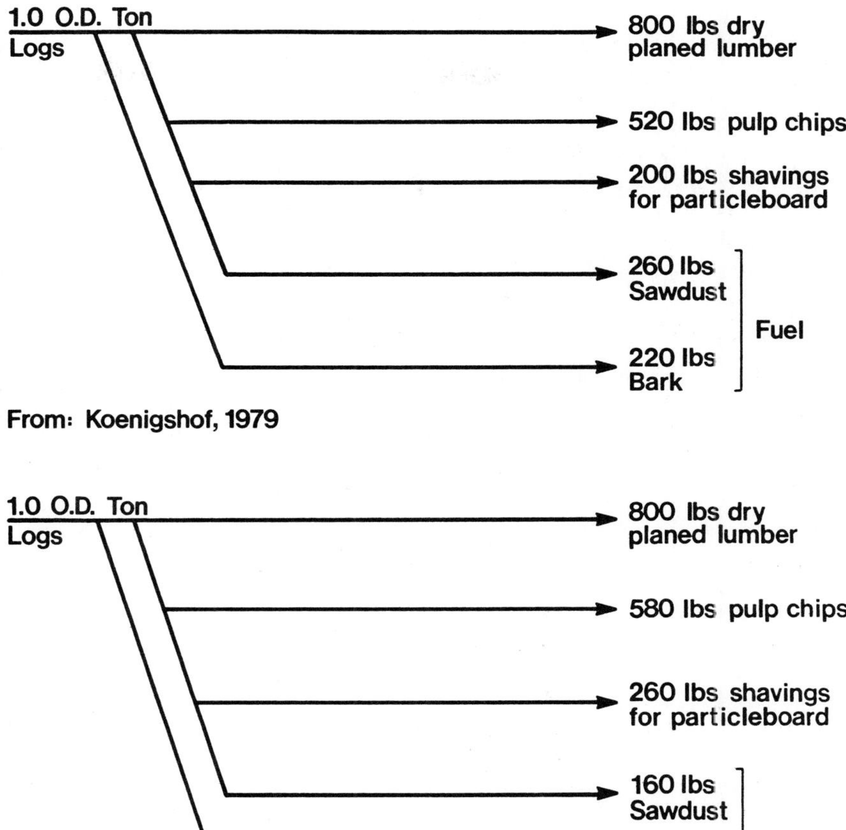

*Figure 10. Yields from the current sawmill as esti-
mated by Koenigshof [20] and CORRIM [6].*

TABLE III. *Lumber and Byproduct Yields for Small Log Sawmills by Mill Type (Basis: Cubic Recovery)*

Mill type	Lumber (%)	Bark (%)	Sawdust (%)	Pulp chips (%)	Shavings and trim (%)
Circular and cant gang	55	10	17	12	6
Band and cant gang	54	10	13	18	5
Band and sash gang	54	10	11	20	5
Chipping headrig	50	10	4	31	5
Scrag	45	10	18	22	6
Log gang	50	10	11	23	6

Source: Adapted from Hartman et al. [16].

individual mill design. At the same time the yields posited in Table III are necessary for overview economic analysis and are therefore used here.

3. Energy Consumption in the State-of-the-Art Sawmill. Energy has been identified as a major cost issue in the production of lumber products. Boyd *et al.* [6] identified the mechanical energy as 180 horsepower hours (hph)/ton of wood. This converts to 210 hph/thousand board feet (MBF), or about 160 kWh/MBF.

Energy for the dry kiln is the other dominant use. Comstock [8] has identified the heat energy requirements by species, as shown in Table IV. The values in Table IV equate to about 1500 lb steam/MBF of Douglas fir, and 3500 lb steam/ MBF of Southern yellow pines.

It is convenient to postulate energy conservation factors for the lumber industry. However, the rising use of chipping headrigs, the advent of scanners and computers, and the trade-off of throughput vs. thermodynamic efficiency in the dry kiln probably mitigates against such factors for the foreseeable future.

TABLE IV.   Lumber Drying Energy Requirements as a
   Function of Wood Species

| | | | Energy use | |
Species	Specific gravity	Water content (lb/ft^3)	Btu/lb H_2O	$1{\times}10^6$ Btu/ MBF
Douglas fir	.45	16.8	2000-3000	1.2-1.8
Southern yellow pines[a]	.47	29.3	1600-2200	3.0-4.0
Red oak	.56	27.9	3000+	6.5+

[a]Loblolly, slash, shortleaf, and longleaf pines.
   Source: Comstock [8].

C.   The Costs of Lumber Production

The costs associated with state-of-the-art sawmilling
have been described by Boyd et al. [6], Envirosphere [39],
and others. Costs as presented here include capital costs
and such operating and maintenance costs as wood, labor,
energy, and other expenses.

1.   Capital Costs. Capital costs have been estimated
for two sized sawmills, a 25×10^6 bd ft/yr mill and a
100×10^6 bd ft/yr mill. The former is based upon the twin
band headrig stud mill of Simpson Timber (Mill #5) as
reported by Williston [42]. The latter is based upon the
data reported by Envirosphere [39].

The capital costs for the two mills selected reported in
both 1980 and (updated) 1984 dollars are shown in Table V.
These capital costs provide an exponential scale factor of
about 0.8 (0.79) for order of magnitude capital cost estimates in sawmills. Between 25 and 100 million bd ft/shift-yr,
then, capital costs may be approximated as follows:

$$\$X = \$11.3 \times 10^6 \left(\frac{C}{25 \times 10^6} \right)^{0.8} \tag{6-1}$$

Where $X is the capital cost of the postulated mill and C is
the annual capacity of the postulated mill in board feet.

2.   Operating and Maintenance Costs. Operating and
maintenance costs can be separated into fixed and variable
costs. Typically labor, maintenance, and overhead are the

TABLE V.   Capital Costs of Sawmills as a Function of System Size

	Capital Cost	
	1980$	*1984$*
25 × 10⁶ bd ft/yr (100 MBF/8 hr shift)	8.7×10^6	11.3×10^6
100 × 10⁶ bd ft/yr (400 MBF/8 hr shift)	26.0×10^6	33.8×10^6

Sources: Williston [42]; Tillman, Rossi, and Simmons [39].

primary fixed O & M costs while raw material and energy are
the dominant variable costs. Labor requirements have been
estimated to be nine workers/shift at the Simpson mill, equal
to $225,000/yr for a 25×10^6 bd ft/yr operation. This in-
cludes a worker earning $8/hr [40] and a fringe benefit
multiplier of 1.5 to achieve a fully loaded wage rate.

Maintenance costs are estimated at 3% of capital costs.
This yields an annual cost of $340,000 for the smaller mill
and $1,010,000 for the larger mill. Supervision and over-
head (including the cost of selling) represents the expenditure of some $250,000 [39] in 1984 dollars.

Wood costs can be calculated from stumpage prices of 1981
as updated. Such prices are as follows [40]: Douglas-fir,
$350/MBF in 1981 or $1.32/ft^3 in 1984; and Southern Pines,
$172/MBF in 1981, or $0.65/ft^3 in 1984. These stumpage
prices have been converted into annual costs as shown in
Table VI.

Energy costs are readily established from the energy
flows previously discussed. Such costs are $12.60-$16.60/MBF
depending upon region, and vary largely as a function of the
quantity of water that must be removed. These energy costs
are summarized in Table VII. All operating costs are summarized in Table VIII.

D. Cost/Price Squeeze in Lumber Production

There is a serious cost/price squeeze in lumber production caused by overcapacity and intermaterial competition as
discussed in Chapter I. The overwhelming need is to create a
more marketable product by reducing raw material costs
through improved yields. With raw materials (timber)

TABLE VI. *Annual Wood Costs for Sawmills in the Northwest and Southeast*

Mill size	Wood costs[a]	
	Douglas-fir	Southern pine
25×10^6 bd ft/yr	3.83×10^6	1.89×10^6
100×10^6 bd ft/yr	15.31×10^6	7.54×10^6

[a]*Based upon 1981 stumpage values published in Statistical Abstract, 1984 [40] as updated.*

TABLE VII. *Annual Energy Costs for Lumber Production*

Mill	Energy cost		
	Electricity	Heat[a] (@2×10^6 Btu)	Total
Southern Pines	(6¢/kWh)		
25×10^6 bd ft	$240,000	$250,000	$490,000 ($16.6/MBF)
100×10^6 bd ft	$960,000	$1,000,000	$1,960,000 ($19.6/MBF)
Northwestern	(4¢/kWh)		
25×10^6 bd ft	$160,000	$107,000	$267,000 ($10.7/MBF)
100×10^6 bd ft	$640,000	$429,000	$1,069,000 ($10.7/MBF)

[a]*Boilers operating @ 70% thermal efficiency.*

consuming 60-84% of all annual costs, it is readily apparent why major efforts have been designed to improve lumber yields. Some of those efforts are described below.

TABLE VIII. Annual Operating Costs for Lumber Production (Not including byproduct credits) (1984 $ \times 10^3$)

| | Region | | | |
| | Southeast | | Northwest | |
Cost	Dollars	%	Dollars	%
25×10^6 bd ft/yr				
Labor	225	7.0	225	4.6
Maintenance	340	10.6	340	6.9
Overhead	250	7.8	250	5.1
Wood	1,890	59.2	3,830	78.0
Energy	490	15.4	267	5.4
Total	3,195	100.0	4,912	100.0
Cost/MBF	128		196	
100×10^6 bd ft/yr				
Labor	375	3.4	375	2.1
Maintenance	1,010	9.1	1,010	5.6
Overhead	250	2.2	250	1.4
Wood	7,540	67.7	15,310	85.0
Energy	1,960	17.6	1,069	5.9
Total	11,135	100.0	18,014	100.0
Cost/MBF	111		180	

Sources: [28, 39, 40, 42].

III. POTENTIAL INNOVATION TO IMPROVE LUMBER PRODUCTION

Potential innovations in lumber production being proposed
for adoption between 1985 and 2000 include improvements at
the individual machine centers of the sawmill and alternatives
to the conventional lumber production system. Innovations
for individual machine centers have been focused upon the
headrig, the dry kiln, and the information/control system.
Proposed alternatives to the conventional sawmill include
sectorwood, wide use of the laminated veneer lumber (LVL)
process, and composite based (Com-Ply) lumber. All of these
potential innovations are surveyed briefly below.

A. Innovations for Machine Center Operations

Innovations for individual machine centers have been proposed continuously based upon the high priority for improving
sawmill yield, and the fact that each individual cut in the
log or cant decreases the degrees of freedom for all subsequent cuts. Dry kiln innovations were advanced most seriously during the period 1972-1980, when the price of oil increased from \$6/bbl to \$40/bbl and there was a consequent
emphasis upon this cost component. Advances in information
and control systems were largely an outgrowth of the computer
hardware and software revolutions, and followed the development of numerical controls in the machine tool industry.

1. Innovations in Headrigs. Because of the need for
yield improvements, most of the innovations proposed for cutting have been focused on the headrig. Innovations proposed
in this area have included the following: (1) the shaping
lathe headrig, (2) slicing headrigs, (3) water jet headrigs,
and (4) laser headrigs.

a. The shaping lathe headrig. The shaping lathe headrig
is perhaps closest to current practice, in that it is an extension of chipping headrig technology. It was developed to
increase the use of hardwoods, and can be integrated with a
structural flakeboard plant [19].
The shaping lathe headrig handles logs of 6 to 9 ft in
length and 6 to 13 in. diameter inside bark (DIB) at the
small end. By use of cams and knives operating parallel to
the grain, the shaping lathe headrig can produce square,
hexagonal, octagonal, or cyclindrical shaped cants and bolts.
Its products can be used as railroad ties, posts and poles,
fence rails, pallet and container shook, flooring, furniture
stock, and a host of other products [19]. Those are the

products envisioned from hardwoods; however there is no inherent technical reason why this process could not be applied to softwoods.

The shaping lathe headrig based mill is not a high yield operation. The typical material balance is shown in Fig. 11. Figure 11 illustrates that its lumber yield from a mill based on this headrig is about 38% and chip yield is about 34% [19]. Such values are comparable to average (but not state-of-the art) chipping headrig sawmills today. The shaping lathe system can produce intricate shapes close to final product requirements (e.g., bolts for hexagonal flooring blocks), however. Further it produces higher quality (more uniform and smooth) chips than the existing chipping headrigs [41]. These characteristics indicate utility as a specialty oriented headrig for this particular innovation.

b. Slicing headrigs. Slicing headrigs offer a direct approach to increasing the yield of the headrig by adapting the veneer lathe to lumber production. This development is a logical extension of efforts to slice thick veneers.

In the slicing approach, logs are heated and then sliced on a knife backed up by a pressure roller bar. Flitches with thicknesses of up to one inch have been cut in this manner,

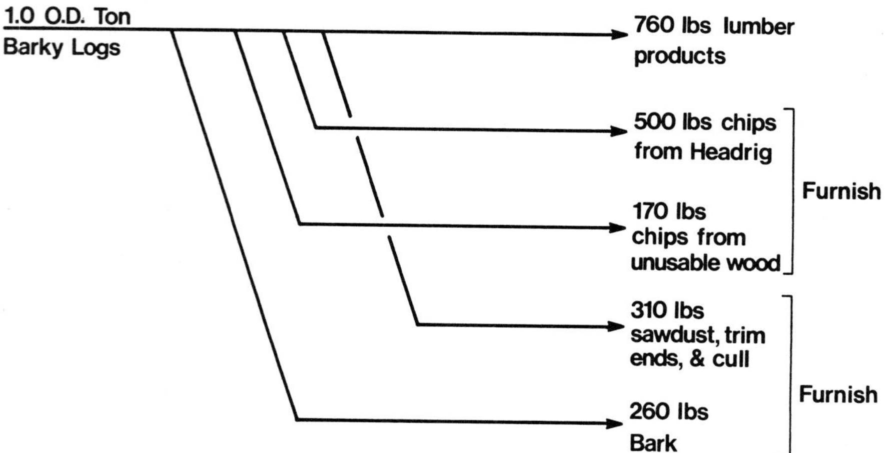

FIGURE 11. *Yields from the shaping lathe headrig [19].*

although knife side fractures in the flitch become more
severe when relatively thick pieces are cut. Knife side
fracture problems are shown in Table IX. Cutting rates of
slicers have approached 5 ft/min with compressions of up to
20% [31]. This contrasts with the conventional chipping
headrig feed rate of 150 ft/min [1]. Higher feed rates of
wood into slicing headrigs could cause hydraulic rupture of
the wood. Consequently feed rates above 5 ft/min have not
been attempted for this primary breakdown system. Similarly
lower feed rates (as slow as 3 ft/hr) have been considered
but such feed rates have not reduced knife side fracture
problems [31]. Because of thickness limitations, fracture,
and feed rate problems, the potential innovation of slicing
headrigs appears to be of limited value.

   c. *Water jet and laser cutting.* Water jet and carbon
dioxide laser cutting are based upon technology transfer from
other industries to the FPI [37]. They offer, at best,
specialty application potential.
   Water jet cutting (or liquid polymer jet cutting) is
based upon developments in mining and metallurgy. It has
been proposed as a means for kerf reduction. Like slicing,
however, liquid jet cutting is severely limited in terms of
feed rate. Further its energy consumption per in.^{3} of material removed was 17.3 times that associated with a conventional bandmill. Kerf losses can be reduced by a factor of
6 by switching from conventional saws to liquid jet cutting.
Therefore the net energy consumption per lineal foot of
material manufactured is increased by a factor of 3. Liquid
jet cutting would slow the entire mill down while increasing
its energy requirement. Water jet cutting has been used successfully in paper cutting where the compressive strength of
the material being cut is not sufficient to reduce operating
rates [37, 39].

*TABLE IX. Knife Side Fracture Depths on Slicing
   Headrigs as a Function of Flitch Thickness*

Flitch thickness	Knife side fraction depth (% of flitch thickness)
1/4 inch	0 - 10%
1/2 inch	10 - 50%
1 inch	30 - 80%

Source: Peters [31].

Laser cutting suffers from the same problem associated with liquid jet cutting. It is a slow speed (low throughput), energy intensive process. Laser cutting feed rates have been measured at 1-1.5 ft/min using a single laser instead of a conventional saw [37]. More recently speeds of 33 ft/min have been achieved [26]. The higher feed rate is based upon two lasers, one making the top half of the cut and the other sending a beam to penetrate the bottom half of the material to be sawn. In comparison, the chipping headrig mill has a feed rate of 9 logs 16 ft long/min, or 150 ft/min [1]. The conventional bandmill can have a feed rate of 96 ft/min [1]. Specific energy requirements for a 240 W carbon dioxide laser approach 100 kWh/in^3 of material removed compared to an energy expenditure of 0.5 kWh/in^3 removal for conventional technologies. As a consequence laser cutting increases both the capital and energy cost components of the sawmill. Kerf losses may or may not be reduced depending upon the extent of efforts made to overcome slow feed rates [39]. Laser cutting is highly suited to careful execution of intricate shape cutting (e.g., in some furniture manufacturing operations). Laser cutting is not suited to high volume common lumber production operations due to its feed rate and energy consumption problems, however. Further, it reduces planing needs.

2. *Innovations in Drying Technology.* Kiln drying originally emerged as a means for reducing the costs associated with work in process, and the related costs of holding inventory in the sawmill. Over time, incremental improvements were made in kiln design (e.g., the CRT kiln). Steam pressures used in the dry kiln increased from 15 psig to 150 psig as new high temperature kiln designs increased drying rates and unit throughputs. These incremental improvements decreased the capital charges per thousand board feet (MBF) of lumber dried. Ultimately the tradeoff was reached between drying rate and degrade generation.

In recent years two alternative drying systems have been promoted aggressively: (1) microwave drying, and (2) solar drying. Both technologies are reviewed below in cursory fashion.

a. *Microwave drying.* Microwave drying originally was proposed in Europe in the 1930s, and then was revitalized by MacMillan Bloedel Corporation as a potential alternative to conventional systems. It was found that microwave systems could increase the rate of drying for individual pieces of lumber dramatically. Rates of drying for large scale tests achieved 11.2%/hr moisture content reduction for Western

Hemlock and 7.1% moisture content reduction for Douglas fir.
Degrade was minimized [3].

Microwave drying, as envisioned by MacMillan Bloedel,
involved using a 25 kW microwave generating unit in order to
achieve reasonable (76%) electrical energy efficiencies. Each
piece of lumber was treated individually. The wave guide
system was a single hybrid ring design oriented at a 40 degree
angle to the flow of lumber [3]. The system was a technical,
but not economic, success [3]. An analogous system has been
developed: the radio frequency/vacuum (RF/V) wood drying
method. This system has been tested successfully in a furniture mill [15], and may have some specialty applications.

The microwave unit is no more cost effective than the
conventional kiln when a cost of capital of 16% was required
[3]. In 1976 dollars, conventional drying costs were
reported to be $12-15/MBF (equal to $25-31/MBF in 1984 dollars) compared to $51-74/MBF for microwave drying under com-
parable conditions (or $107-155/MBF in 1984 dollars) based
upon Barnes *et al.* [3]. Like laser cutting and the RF/V
system, microwave drying appears best suited to special
situations.

b. Solar drying. Solar drying, and solar assisted drying, also have been proposed as alternative drying systems.
Such systems are discussed completely in Smith [32]. Both
greenhouse and solar collector based kilns have been proposed.

Like the microwave alternative, the solar drying system
is sufficiently expensive that it probably will not be used
broadly. It is inherently rate limited. The problem of
production rate leads to high capital charges ($/MBF dried)
and high inventory costs associated with work in process.
Relative production rates are shown in Table X. Solar drying,
like other processes, will probably be used only in special
situations.

*TABLE X. Relative Drying Rates for Lumber by Type
of Dry Kiln*

Drying system	Relative drying rate
Conventional	1.0
Solar collector	2.7 - 17.7
Greenhouse	10.0 - 20.0

Source: Tillman, Rossi, and Simmons [39].

3. Instrumentation and Control. While many proposed
innovations in such machine centers as headrigs and dry kilns
have been unpromising for high production systems, the revolution in instrumentation and controls has been phenomenal.
Innovations in this area are discussed in Comstock *et al.*[9],
Harke [13]; Neild [29]; Funt [11]; Nelson [30]; McMillin,
Conners, and Huber [26]; Bowering [5]; Williston [41]; and
elsewhere.

Innovations in this area that have been proposed for
future adoption include:

(1) 3-dimensional CT scanning of logs both at the merchandising and headrig centers in order to minimize
yield losses due to knots and defect;

(2) microprocessor control of all machine centers, including the dry kiln; and

(3) computer connection of all microprocessors in order
to communicate process problems to all machines down-
stream and upstream of the point of difficulty.

Because the area of instrumentation and control holds significant potential, one of its advanced concepts is used as the
focus of detailed analysis in this chapter.

4. Unit Operations Conclusions. Improvements in specific machine centers have been focused more on special situations than general high production machines. The only excep-
tion is in the area of instrumentation and control.

Cutting technologies will not change significantly from
present methods, at least in the production of dimension
lumber. Novel cutting methods, originally designed for use
in other industries, may have specialty application (e.g.,
the cutting of intricate shapes). They are not appropriate
or economic for most lumber production applications. Novel
drying technologies, like cutting technologies, do not offer
economic advantages when compared to conventional systems.
They do have special applications in certain situations
(e.g., where degrade minimization is paramount). Consequently their use will be limited.

B. Alternatives to the Conventional Sawmill

Three alternatives to the conventional sawmill have been
proposed: (1) sectorwood, (2) laminated veneer lumber (LVL),
and (3) com-ply production. Each of these alternatives
offers a unique approach to improving sawmill yield, and consequently the reduction in raw material cost. These three

approaches to lumber production are more dramatic departures from conventional sawmilling than the unit operations innovations discussed above.

1. *Sectorwood.* The sectorwood system is most similar to the conventional sawmill of the alternatives listed. The process involves debarking the log, splitting it into six or eight pie shaped pieces, drying the pieces, gluing together two pieces into a cant, and then processing the lumber conventionally. The key to the operation is the splitting pattern and the subsequent cant shape.

One can calculate yields of 60 to 65% with sectorwood applied to small logs. This increase in yield is based upon decreased losses at the resaw. Such yields would be 15 to 25% greater than current state of the art in sawmilling. Sufficient problems exist in drying, and associated drying stresses, to make this system difficult to employ. Further, log problems including sweep make sectorwood less attractive than initially believed.

2. *Laminated Veneer Lumber.* The production of LVL is a more radical departure, bringing plywood technology to the production of beams, truss materials, wide boards and even studs. The following sequence of operations is used in LVL production: (1) debarking, (2) merchandising and bucking to length, (3) lathe charging, (4) peeling, (5) veneer drying, (6) veneer layup, (7) gluing (8) pressing, and (9) ripping to final dimensions.

The concept of parallel laminated veneer (PVL), upon which LVL is based, began in the 1940s. PLV was viewed as a means for producing high strength aircraft materials. Since that time PLV and LVL have been commercialized for the production of specialty products by Truss Joist Corporation, Weyerhaeuser Company, and Metsaliiton Teollisuus Oy of Finland [23].

The attractive features of LVL stem from the fact that it is more of an engineered material than conventional lumber. It can be produced in virtually any length or width desired, by use of continuous pressing. It can be produced at high strengths. Joint strengths can be varied by use of butt and crushed butt joints, lapp joints, scarf joints, and reinforced butt joints [23]. Board strengths can be varied not only as a function of joint type and adhesive, but also as a function of reinforcing agent. LVL can be reinforced with fiberglass or graphite fibers to make high strength boards [24].

Being an engineered material, LVL not only can be manufactured to a variety of strengths, but can also be manufactured to more uniform strengths than conventional lumber.

The MICRO-LAM Ⓡ product of Truss Joist Corporation can be used to illustrate this uniformity characteristic. The tensile testing of MICRO-LAM Ⓡ had a coefficient of variation (COV) of 12%. Machine stress rated lumber has a COV of 25% while visually graded lumber has a COV of 35% [23]. Product strength uniformity is a function of the veneer thickness used in making LVL.

Already LVL has become a highly desirable material for the manufacture of trusses, scaffolding, and other situations where long lengths or wide boards are required. At present some 20 million board feet are being made annually [21]. Improvements are being made in numerous aspects of LVL manufacture, including adhesives [25].

LVL shows great promise for many applications, however it is not likely to be a total substitute for dimension lumber. Its cost structure is limiting, as shown by cost factors presented in Table XI. Given the factors shown in Table XI, it is apparent that LVL provides a useful technique for making engineered wide boards, long boards, truss materials, and similar products. LVL, then, appears to be a useful complement to conventional sawing, and one that will grow in years to come.

TABLE XI. Economic Factors for Laminated Veneer Lumber Relative to Conventional Sawmill Systems

	Relative Factor	
Cost factor	Conventional lumber	LVL[a]
Capital ($/MBF)	1	2.95
Yield (MBF lumber/ton of logs)	1	1.34[b]
Labor (hrs/MBF of product)	1	1.47
Heat energy (Btu/MBF of product)	1	1.48
Electrical energy (kWh/MBF of product)	1	0.17

[a]*Sources: Boyd et al., [6] yields confirmed by Harpole [14].*

[b]*Kunesh [21] attributes the 25-50% increase in yield to elimination of losses in squaring the log and sawdust.*

3. Com-Ply Lumber. Com-ply is the third alternative to
the conventional lumber mill. Com-ply lumber consists of
boards with particleboard cores and veneer skins. Typically
the board is projected to be about 30% 1/4 in. veneer (by
volume) and 70% particleboard. Its manufacture requires not
only sound logs but also provides an outlet for junk trees
[20].

Com-ply lumber has many of the market oriented characteristics associated with LVL. It can be produced at virtually
any length or width. It can be produced to high degrees of
uniformity. These advantages are shown in detail by Bloomquist *et al.* [4]. Such advantages do not necessarily translate into economic gains, however, as is shown in Table XII.
On this basis Com-ply does not appear as promising as
originally anticipated.

C. Conclusions

Of the innovations reviewed in the previous sections,
potential advances in instrumentation and controls appear
most universally appealing. Other innovations within the
machine centers appear most suited to specialty applications
(e.g., the use of lasers to cut intricate patterns). Innovations in the area of alternatives to the conventional sawmill appear most suited to specific markets (e.g., trusses,
scaffolding).

*TABLE XII. Annual Manufacturing Costs for Com-Ply
Compared to Conventional Lumber Production (1984$/MBF)*

	Mill type	
Cost categroy	Com-Ply[a]	Northwest 25×10⁶ BF/yr mill
Wood	67	153
Labor	32	9
Power and fuel	19	11
Resin, wax and adhesives	64	0
Overhead	45	23
Total	227	196

[a] *Sources: [4, 6, 20, 39].*

The advanced controls sawmill, therefore, is selected for detailed analysis. It provides two benefits: (1) an image of the future of lumber production, and (2) an example of technique application in new technology analysis.

IV. ANALYSIS OF THE ADVANCED CONTROLS SAWMILL

The advanced controls sawmill has a very high probability of adoption. This analysis of such a mill takes the form of a brief process description followed by the engineering economy evaluation.

A. Process Description

The advanced controls sawmill has been described conceptually in several publications (e.g., [33, 39]). Its details have been published by such authors as McMillin, Connors, and Huber [26], Baldwin [2], Napier [27], Armstrong [1], Faggard [10], and numerous other authors. Funt [11] offered additional insights in an unpublished paper.
In the advanced controls mill, logs are sorted according to narrow size classifications [42]. Logs then are scanned prior to bucking, using axial tomography [26]. Tomography provides 3-dimensional scanning, such that internal knots and defect can be "seen" by the sawyer. Early wood and late wood also can be detected by tomography. Similarly the interface between heartwood and sapwood can be viewed by this technique. Of equal importance is the fact that tomography can determine the complete geometry (including sweep) of the log.
The current research in this area includes the use of CT scanners, essentially identical to those used in medicine for brain scans [11]. The problems in this area are to increase the feed rates of logs through such tomographic scanners.
Once the log has been scanned, the results are passed through such computer programs as BOF to maximize quantitative and qualitative yield. Such programs determine log bucking. Once the log is bucked, it is passed through a second tomographic scan and computer prior to entering the chipping headrig. The second scan determines the actual primary breakdown pattern. A chipping headrig is probable because of its high rates of throughput. Throughput rates largely determine capital utilization rates [11].
Actual unit operations in this mill are quite similar to the conventional state-of-the-art mill. The edgers and trimmers are microprocessor controlled. The planer uses abrasive planing for loss reduction. Microprocessor

controls and sensors are added to the plane along with the
dry kiln.

The primary difference between this advanced controls
mill and the conventional state-of-the-art sawmill is its
level of intelligence. Not only can the entire log be "seen"
but the individual machine centers can communicate with each
other. The headrig can "tell" the edger and trimmer what is
coming. Similarly, and perhaps more importantly, planers,
trimmers, and edgers can notify upstream machine centers when
significant problems become chronic.

This knowledge system results in improved lumber yield, as
is shown in Fig. 12. The yield from this system is 56%
(mass basis) compared to 50% for the conventional chipping
headrig mill (see Table III). While the process is analyzed
here using a chipping headrig, it applies equally to a bandmill. The yield improvement is estimated at 10-15% [39].

B. Engineering Economy Analysis

In order to perform an engineering economy analysis of
this innovation, it is necessary to analyze costs and revenues. Production costs for a southeastern mill are shown
in Table XIII. These are somewhat at variance to the production costs associated with a northwestern mill due in large
measure to much higher stumpage costs in the northwest (see
Table VI).

Revenues are analyzed in Table XIV. It is useful to note
that there is a net gain of $632,000 in annual revenue for
the 25×10^6 BF/shift-yr (nominal) mill. The increased lumber recovery overwhelms the loss of $118,000/yr in pulp chip
sales. Those lost pulp chips represent an opportunity cost
to the mill owner. It is essential to note, however, that
these revenue estimates assume the sale of all lumber produced, which may or may not be reasonable given the current
overcapacity in the sawmill sector of the FPI.

Discount rates, like costs and revenues, are of primary
importance. On a total system basis, this innovation is
clearly reversible. Consequently the nominal discount rate
is 19% and the real discount rate is 10%. For the innovation
itself, however, the pioneering nominal discount rate of 25%
(real discount rate of 19%) applies.

Total system analysis for a southeastern mill is shown in
Table XV. Incremental analysis for a southeastern mill is
shown in Table XVI. It is useful to note that the addition
of advanced controls in the southeast does not reduce the
levelized price of lumber on a total basis. The gains from
maturity are $5/MBF, but the risk adjusted costs of the in-
crement of production are $260/MBF. The mature costs of the

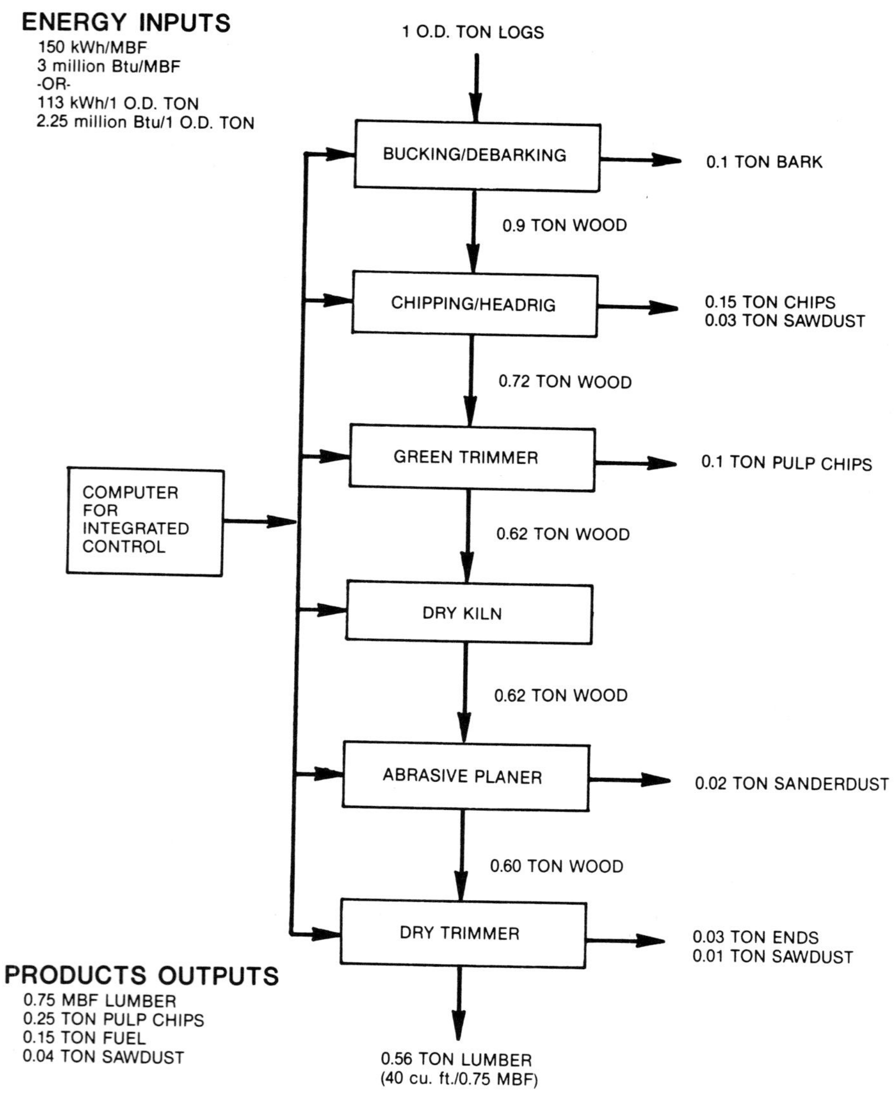

FIGURE 12. *Unit operations and yields from the Advanced Controls sawmill [39].*

TABLE XIII. Cost Analysis for the Nominal 25×10^6 Board Feet/Year Advanced Controls Sawmill in the Southeast (Basis: $1984\$ \times 10^3$)

Parameter	Conventional mills	Advanced controls mill	Difference
Yield (%)			
(Assumes chipping headrig)	*50%*	*56%*	*12%*
Yield (Board feet)	25×10^6	28×10^6	3×10^6
Capital cost	*$11,300*	*$12,960*	*$1,660[a]*
Operating cost			
Labor	*$ 225*	*$ 225*	*0*
Maintenance	*340*	*390*	*$50*
Overhead	*250*	*250*	*0*
Wood	*1,890*	*1,890*	*0*
Energy	*490*	*550*	*60*
Operating cost ($/MBF)	*128*	*118*	*37*
Fixed ($/MBF)	*33 ($815/yr)*	*31 ($865/yr)*	*17 ($50/yr)*
Variable ($/MBF)	*95*	*87*	*20*

[a] Covers tomographic scanner, central computer for communication, and related systems. Assumes existence of currently available microprocessors and related systems.

increment of production caused by advanced knowledge and controls are $210/MBF, just 5% lower than the costs of the current mill.

What becomes interesting is the comparison of dominant costs. Such a comparison is presented in Table XVII. When it is all netted out, the purchase of an advanced controls system in the southeast is about analogous to the purchase of an added increment of capacity—on a mature technology basis.

The advanced controls system also has been analyzed for the northwest. The northwest is characterized by higher stumpage costs and lower energy costs. The advanced controls sawmill shows dramatic promise in the northwest, as is shown in Tables XVIII and XIX. The difference between the southeastern and northwestern cases is in the variable costs.

TABLE XIV. Revenue Analysis for the Nominal 25×10^6 Board Feet/Year Mill (Basis: 1984$)

Assumed unit values

Lumber	*$250/MBF*[a,b]
Pulp chips	*$45/O.D. ton*
Fuel	*$35/O.D. ton ($2/Btu $\times 10^6$)*

Annual revenue analysis
(1984$ $\times 10^3$)

Revenue	Conventional mill	Advanced controls mill	Difference
Lumber	*6,250*	*7,000*	*750*
Pulp chips	*607*[c]	*489*[d]	*(118)*[e]
Fuel	*289*[c]	*289*[d]	*–*
Total	*7,146*	*7,778*	*632*

[a] $1000 \text{ bd ft} \times \dfrac{58 \text{ ft}^3}{MBF} \times \dfrac{30 \text{ lb}}{\text{ft}^3} \times \dfrac{\text{ton}}{2000 \text{ lb}} = 0.87 \text{ ton/MBF}$

Mill takes in 50×10^3 MBF/yr or 43.5×10^3 ton/yr O.D.

[b] *Sources: [28, 40] as updated.*

[c] *This equates to a byproduct credit of $36/MBF.*

[d] *This equates to a byproduct credit of $28/MBF.*

[e] *This equates to a byproduct opportunity cost of $39/MBF.*

Variable costs, including $39/MBF in opportunity costs, total $53/MBF for the advanced controls increment when the mill is located in the northwest. For the conventional mill in the northwest, incremental variable costs total $163/MBF.

It is apparent that advanced controls, including axial tomography and machine center communication, are highly attractive advances in sawmilling for northwestern mills if supported by sufficient lumber demand. Further, these mills using tomography will find "high grading" of the log more readily accomplished, and "high grading" may be absolutely essential in periods of slack lumber demand. Advanced controls are particularly attractive when stumpage costs are

TABLE XV.   Total Analysis of Alternative Systems, Southeastern Mills   (Basis: 1984$ $\times 10^3$)

Parameter	Conventional system	Advanced system risk adjusted	Mature
Costs			
Capital ($)	11,300	12,960	
Operating			
Fixed ($/yr $\times 10^3$)	815	865	
Variable ($)	95/MBF	87/MBF	
Lumber yield	25×10^6 BF	28×10^6 BF	
Revenues ($/yr $\times 10^3$)	7,146	7,778	
Financial values			
Project life (yrs)	20	20	
Tax rate (%)	46	46	
Discount rate (%)			
Nominal	15	19	15
Real	10	13	10
Inflation rate (%)	5	5	5
Net present value			
Nominal	12,290	9,920	13,810
Real	11,640	9,220	13,080
Levelized lumber price (Net of byproduct credits)[a]			
Nominal basis ($)	220[130]	230	220[126]

[a]*Variable costs are $59/MBF for both cases.*

TABLE XVI.   Incremental Analysis of Control System, Southeastern Mill (Basis: 1984$ × 10^3)

Parameter	Value risk adjusted	Mature
Costs		
Capital ($)	*1,660*	*1,660*
Operating and Maintenance		
Fixed ($)	*50*	*50*
Variable ($)	*20/MBF*	*20/MBF*
Lumber yield	3×10^6 *BF*	3×10^6 *BF*
Revenues ($/yr × 10^3)	*632*	*632*
Financial value		
Project life (yrs)	*20*	*20*
Tax rate (%)	*46*	*46*
Nominal discount rate (%)	*25*	*19*
Inflation rate (%)	*5*	*5*
Net present value ($)	*380*	*1,520*
Levelized lumber price (Net of byproduct credits/ costs. Variable cost - $59/MBF)		
Nominal basis ($)	*260*	*210*

high. Consequently such controls will become increasingly important throughout the nation as silvicultural management becomes more important, and raises the cost of stumpage in regions outside the northwest.

V. CONCLUSION

Sawmilling, one of the basic forest industry systems, has experienced significant improvements in recent decades. Conventional milling has experienced gains in numerous specific unit operations. Among these are the development of chipping headrigs, 2-dimensional scanning, and microprocessor controls.

TABLE XVII. Incremental Analysis Values for the Advanced Controls Sawmill Installed in the Southeast

Analysis parameter	Conventional mill value	Controls based production increment
Capacity	25×10^6 BF/yr	3×10^6 BF/yr
Capital cost	11.3×10^6	1.66×10^6
Capital cost/MBF	450	550
Capital cost/MBF net of investment tax credits	410	500
Variable cost ($/MBF)		
Production	59[a]	20
Opportunity	–	39
Total	59	59
Fixed operating cost ($/MBF/yr)	33	17

[a]*Net of byproduct credits.*

In addition to advances in conventional sawmilling, alternatives have emerged or been proposed. These alternatives include LVL and Com-ply processes. Such alternatives are particularly appealing for special applications.

Conventional milling appears to be the dominant process that will be deployed in most cases for the foreseeable future. Within that context numerous improvements have been proposed. Advanced knowledge and controls are the most promising of these alternatives. Such improvements are particularly attractive where stumpage costs are relatively high. Such areas now are focused in the Pacific Northwest, but could extend to any area where silvicultural activities increase stumpage prices.

Advanced controls promise substantial decreases in lumber costs as is highlighted by incremental analysis in the Pacific Northwest. Such decreases are critical. The tremendous overcapacity in sawmilling, and the competition from metallic and plastics products remain as significant problems. Survival in such an environment depends upon becoming the lowest cost producer of lumber.

TABLE XVIII. *Total Analysis of Alternative Systems, Northwestern Mills (Basis: 1984$ × 10^3)*

Parameter	Conventional system	Advanced system risk adjusted	Mature
Costs			
Capital ($)	11,300	12,960	
Operating			
Fixed ($)	815/yr	865/yr	
Variable ($)	163/MBF[127]	147/MBF[119]	
Lumber yield	25 × 10^6 BF	28 × 10^6 BF	
Revenues ($/yr × 10^3)	7,146	7,778	
Financial values			
Project life (yrs)	20	20	
Tax rate (%)	46	46	
Discount rate			
Nominal (%)	15	19	15
Real (%)	10	13	10
Inflation rate (%)	5	5	
Net present value			
Nominal ($)	4,100	3,240	5,770
Real ($)	3,720	2,770	5,290
Levelized lumber price (Net of byproduct credits)[a,b]			
Nominal basis	310	305	300

[a] $36 for conventional system.

[b] $28/MBF for advanced system.

TABLE XIX. *Incremental Analysis of Control System, Northwestern Mill (Basis: 1984\$ $\times$ 10^3)*

	Value	
Parameter	Risk adjusted	Mature
Costs		
Capital (\$)	1,660	1,660
Operating and Maintenance		
Fixed (\$)	50/yr	50/yr
Variable (\$) (production only)	14/MBF	14/MBF
Lumber yield	3×10^6 BF	3×10^6 BF
Revenues (\$)	632	632
Financial values		
Project life (yrs)	20	20
Tax rate (%)	46	46
Nominal discount rate (%)	25	15
Inflation (%)	5	5
Net present value	360	1,440
Levelized lumber price (Net of byproduct credits/ costs)		
Nominal basis	260	200

REFERENCES

1. Armstrong, R. 1984. Three Approaches to High Recovery,
   Small Log Breakdown. *In* "Computer Automation for Saw-
   mill Profit," Proc. Forest Products Research Society,
   Madison, Wisconsin.

2. Baldwin, R. F. 1984. Sawmill Control—What Will the
   Future Bring. *In* "Computer Automation for Sawmill
   Profit," Proc. Forest Products Research Society,
   Madison, Wisconsin.

3. Barnes, D., Admiraal, L., Pike, R., and Mathur, V. 1976.
   Continuous System for the Drying of Lumber With Microwave Energy. *Forest Products Journal 26*(5):31-42.

4. Bloomquist, R. F., Duff, J. E., McAlister, R. H., Fick,
   C. B., Wittenberg, D. C., and Walther, R. M. 1975.
   Com-Ply Status Report. *Forest Products Journal 25*(9):34-
   43.

5. Bowering, C. G. 1980. One Mill's Experience With
   Computer Scanning on a Canter Quad. *In* "35th Annual
   Northwest Wood Products Clinic," Proc. Spokane, Washing-
   ton, May 5-7.

6. Boyd, C. W., Koch, P., McKean, H. B., Morschauser, C. R.,
   Preston, S. B., and Wangaard, F. F. 1976. Wood for
   Structural and Architectural Purposes. *Wood and Fiber
   8*(1):1-72.

7. Brink, D. L., and Merriman, M. M. 1978. Development of
   Bonding Systems Using Modification of the Lignocellulosic
   Matrix and Crosslinking Agents. University of California
   Forest Products Laboratory, Richmond, California

8. Comstock, G. L. 1976. Energy Requirements for Drying of
   Wood Products. *In* "Wood Residue As an Energy Source,"
   Proc. Forest Products Research Society, Madison,
   Wisconsin.

9. Comstock, G. L., Adams, W. L., Gravely, E. G., Lamb, F.M.,
   Nearn, W. T., Noffsinger, J. R., and White, M. S. 1984.
   Computer Automation for Sawmill Profit. Forest Products
   Research Society, Madison, Wisconsin.

10. Faggard, D. W. 1984. Process Controls—Programmable
    Controller vs Computer. *In* "Computer Automation for
    Sawmill Profit," Proc. Forest Products Research Society,
    Madison, Wisconsin.

11. Funt, B.   CT Scanning of Logs.   Presented at 33th Annual
Meeting, Forest Products Research Society, St. Louis,
June 27.

12. Hand, D. E.   1979.   Log Merchandizing.   *In* "34th Annual
Northwest Wood Products Clinic," Proc.   Kalispell,
Montana, April 30-May 2.

13. Harke, S. R.   1984.   Process Control in the Forest
Products Industry: Past and Present.   Presented at the
38th Annual Meeting, Forest Products Research Society,
St. Louis, June 27.

14. Harpole, G. B.   1976.   Yield Comparison: Press-lam vs.
Sawn Lumber and Plywood.   *Forest Industries* 103(9):42-44.

15. Harris, R. A., Taras, M. A., and Schroeder, J. G.   1984.
Sound Quality Upholstered Frame Part Yields From Lumber
and Green Cuttings Dried by a Radio-Frequency/vacuum
System and by Conventional Kiln-Drying.   *Forest Products
Journal* 34(7/8):19-21.

16. Hartman, D. A., Atkinson, W. B., Bryant, B. S., and
Woodfin, R. O.   1978.   Conversion Factors for the Pacific
Northwest Forest Industry.   University of Washington,
Seattle, Washington.

17. Haygreen, J. G., and Bowyer, J. L.   1982. "Forest Products and Wood Science: An Introduction."   Iowa State
University Press, Ames, Iowa.

18. Judson, P.   1979.   Tilted Headrigs.   *In* "34th Annual
Northwest Wood Products Clinic."   Kalispell, Montana,
Apr. 30-May 2.

19. Koch, P., and Caughey, R. A.   1978.   Shaping-Lathe
Headrig Yields Solid and Molded-Flake Hardwood Products.
Forest Products Journal 28(10):53-61.   Also see
Springate, N., Plough, I., and Koch, P.   1978.   Shaping-
Lathe Roundup Machine Is Key to Profitable Manufacture
of Composite Sheathing Panels in Massachusetts or Maine.
Forest Products Journal 28(10):42-47.

20. Koenigshof, G. A.   1979.   Status of Com-Ply Floor Joist
Research.   *Forest Products Journal* 29(1):37-42.

21. Kunesh, R. H.   1978.   Micro-Lam Ⓡ : Structural Laminated
Veneer Lumber.   *Forest Products Journal* 28(7):41-44.

22. Kuo, M. L., Birnbach, M., Brink, D. L., Johns, W. E.,
Layton, D., Nguyen, T., and Wong, A. Development of Bond-
ing Systems Using Modification of the Lignocellulosic
Matris and Crosslinking Agents.   University of California
Forest Products Laboratory, Richmond, California.

23. Laufenberg, T. L. 1983. Parallel-laminated Veneer:
Processing and Performance Research Review. *Forest
Products Journal 33*(9):21-28.

24. Laufenberg, T. L., Rowlands, R. E., and Krueger, G. P.
1984. Economic Feasibility of Synthetic Fiber Reinforced
Laminated Veneer Lumber (LVL). *Forest Products Journal
34*(4):15-22.

25. Loehnertz, S.P. 1983. Cost Comparison of Two Processes
for Laminating Thick Veneer. *Forest Products Journal
33*(11/12):57-60.

26. McMillin, C. W., Conners, R. W., and Huber, H. A. 1984.
ALPS—A Potential New Automated Lumber Processing System.
Forest Products Journal 34(1):13-20.

27. Napier, T. W. 1984. Implementation of Process Control
Technology. *In* "Computer Automation for Sawmill Profit,"
Proc. Forest Products Research Society, Madison,
Wisconsin.

28. National Forest Products Association. 1983. Composite
Prices of Softwood Lumber and Plywood.

29. Neild, P. 1984. Better Sawmill Management Through Computer Simulation. Presented at the 38th Annual Meeting
of the Forest Products Research Society, St. Louis,
June 28.

30. Nelson, W. C. 1984. Small Log Sawmill Systems: Past,
Present, and Future. Presented at the 38th Annual
Meeting of the Forest Products Research Society, St.
Louis, June 27.

31. Peters, C. C. 1976. Basis and Specifications for a
Lumber Slicer. *Forest Products Journal 26*(2):26-30.

32. Smith, W. R. 1980. An Economic Model of the Lumber
Drying System. PhD Dissertation. University of California, Berkeley, California.

33. Smith, W. R. 1982. Personal communication to Envirosphere Co., Jan. 7.

34. Snellgrove, T. A. 1975. Lumber Grade Recovery as
Affected by Different Types of Processing Facilities.
Forest Products Research Society, Madison, Wisconsin.

35. Soltes, E. J., and Lin, S. 1983. Adhesives From
Natural Resources. *In* "Progress in Biomass Conversion,"
Vol. 4. Academic Press, New York.

36. Stone, E. J., and Ott, N. 1984. Log Sorting to Increase
    Profits. *In* "Computer Automation for Sawmill Profit,"
    Proc. Forest Products Research Society, Madison,
    Wisconsin.

37. Szmani, R., and Dickenson, F. E. 1975. Recent Develop-
    ments in Wood Machining Processes: Novel Cutting Tech-
    niques. *Wood Science and Technology* 9(2):113-128.

38. Tillman, D. A., Rossi, A. J., and Kitto, W. D. 1981.
    "Wood Combustion: Principles, Processes, and Economics."
    Academic Press, New York.

39. Tillman, D. A., Rossi, A. J., and Simmons, S. O. 1982.
    Wood: Its Present and Potential Uses. Envirosphere Co.,
    Bellevue, Washington. For the U.S. Congress, Office of
    Technology Assessment.

40. U.S. Department of Commerce. 1984. Statistical Abstract.
    U.S. Government Printing Office, Washington, D.C.

41. Williston, E. 1979. State-of-the-Art Report: Lumber
    Manufacture. *Forest Products Journal* 29(10):45-49.

42. Williston, E. 1981. "Small Log Sawmills." Miller
    Freeman, San Francisco.

Chapter VII

ADVANCED TECHNOLOGIES FOR PRODUCING ENERGY,
CHEMICALS, AND RELATED PRODUCTS FROM WOOD

I. INTRODUCTION

The Forest Products Industry (FPI) derives most of its
wealth from the production of pulp and lumber. At the same
time it produces a variety of energy, chemical, and related
products as shown in Table I. These products generate over
$6 billion in annual revenues for the FPI.

Several features are common to energy, chemical and related products made from wood: (1) they are largely used in
specialized rather than broad (bulk commodity) applications;
(2) they are based upon the use of byproducts from cther
forest industry operations; (3) they are based largely upon
the chemical properties of wood; and (4) they hold more
promise for dramatic growth than the traditional segments of
the industry.[1]

Producing energy, chemicals, and related products is not
new to the FPI. Wood was the primary fuel used in the U.S.
until 1870, and the dominant feedstock for plastics manufacture until 1940 [7, 17]. The chemicals industry began as
the Naval Stores industry. Revival of interest in these
areas resulted from the restructuring of oil and export foodstuff pricing that occurred during the 1970s.

While significant growth potential exists for producing
energy, chemicals, and related products from wood, these
goods will not displace pulp or lumber as the dominant commodities of the FPI in the foreseeable future. Consequently
energy, chemicals and related products are treated as a
single unit in this monograph.

---

[1]*Steam and electricity are bulk commodities, however
their generation with wood fuel has required special situations to date. Similarly whole tree chips are not byproducts
of manufacturing but byproducts of silvicultural activities.*

TABLE I.   Representative Energy, Chemicals, and Specialty Products of the Forest Products Industry

---

Representative energy products

   Whole tree chip fuel

   Hog fuel

   Wood pellets

   Firewood and cordwood

   Sawdust logs and charcoal

   Steam

   Direct heat

   Electricity

Representative chemicals and related products

   Turpentine

   Rosin

   Tall oil

   Cellophane

   Rayon

   Cellulose acetate

   Ethanol

   Dimethyl sulfoxide (DMSO) solvent

Representative specialty products

   Cattle food supplements

   Drilling mud additives

   Vanillan

II. THE STATE OF THE ART IN ENERGY, CHEMICALS, AND RELATED PRODUCT PRODUCTION

Currently wood is used as a feedstock for many energy, chemical, and specialty products (e.g., electricity, turpentine). Within this arena of energy, chemicals, and related products, energy has become the dominant economic contributor to the FPI. Consequently, energy is given the most extensive treatment here.

A. *The State-of-the-Art in Energy Generation*

Wood has been used as a fuel since the discovery and control of fire [38]. Further, such analysts as Glesinger [24] have foretold of its dramatic resurgence. Today wood finds use in industrial and utility boilers as a source cf energy for both process steam and electric power generaticn. Wood is emerging as a fuel for large scale cogeneration systems such as the 30MW/500,000 lb process steam cogeneration system at Longview, Washington, depicted in Fig. 1. Wood also is used to fuel stand alone condensing power plants such as the 50 MW station in Burlington, Vermont, depicted in Fig. 2. Other wood fired power plants completed recently ard now in operation include the 47MW station built by Washington Water Power at Kettle Falls, Washington, and the 15MW station retrofit to burn wood at French Island, LaCrosse, Wisconsin. The 15MW retrofit unit is based upon fluidized bed combustion [72]. Wood also is used to fuel the boilers of the Concord Steam Corp., a steam utility in Concord, New Hampshire [2].

Electricity generation and utility fuel sales have become one of the major wood products growth areas. Wood fired industrial and utility cogeneration and condensing power generating facilities now have a total capacity of about 1,400MW [11]. This capacity can produce about 9.8×10^9 kWh/yr, worth some $500 to $600 $\times 10^6$ annually at 5¢/kWh to 6¢/kWh. Wood fired electricity generation growth is based upon the ability of industry to capture markets created by the Public Utilities Regulatory and Policies Act (PURPA), and the ability of utilities to build wood fired units on accelerated schedules.

Wood fired power plants may obtain their fuel from a variety of sources. Of the specific plants mentioned above, the Longview, Kettle Falls, and French Island plants use residues from local sawmills and pulp mills (mill residuals). The Burlington plant is fueled by forestry residuals. The Burlington plant demonstrates that, in general, the availability of mill residues is limited [14, 25, 64]. Dependable supplies of wood fuels now must come from the forest, as shown in Fig. 3. Forest fuels are relatively expensive. Consequently considerable attention is being given to make wood energy systems more efficient. This emphasis on efficiency is seen in combustor and overall system design.

A second force impacting wood energy is the development of alternative wood energy systems. Such alternatives

FIGURE 1. *The 30MW/500,000 lb process steam wood fired cogeneration facility of Weyerhaeuser at its integrated mill in Longview, Washington. (Photo courtesy of Weyerhaeuser Co.)*

FIGURE 2. *The 50MW wood fired condensing power plant in Burlington, Vermont.*

include gasifiers used to produce low Btu gas. Gasification can be used as an alternative to wood combustion depending upon product or regulatory constraints. These forces lead to consideration of the following three issues: (1) combustor improvements, (2) gasification designs, and (3) general electricity generation system improvements.

1. *Combustion System Improvements.* Direct combustion revolves around driving the following reactions:

$$C + O_2 \longrightarrow CO_2 + 169,290 \text{ Btu} \tag{7-1}$$

$$H_2 + 1/2\ O_2 \longrightarrow H_2O + 122,970 \text{ Btu} \tag{7-2}$$

$$C + 1/2\ O_2 \longrightarrow CO + 47,550 \text{ Btu} \tag{7-3}$$

FIGURE 3. *The production of whole tree chips for fuel use. Forest residuals are now the incremental supply of wood fuel.*

$$CO + 1/2\ O_2 \longrightarrow CO_2 + 121,740\ Btu \qquad\qquad (7\text{-}4)$$

Numerous systems have been designed to capitalize upon these reactions including Dutch ovens, undergrate stokers, flat grate spreader-stokers, and similar devices (see Leppa [4]). Until about 1967 the primary objective of many such systems was as much to incinerate waste wood as to generate useful heat and power. More recently efficient energy recovery has become most important [59].

By 1975, the state-of-the-art in wood combustion was the traveling grate spreader-stoker. Atmospheric fluidized bed combustion (AFBC) had been introduced for special problems. Suspension and cyclonic burners had been developed for special dry fuels such as planer shavings and sanderdust. Inclined grates were being introduced from Europe, and being built by some U.S. manufacturers.

Since 1975 advanced concepts have been developed in such areas as fluidized bed combustion. These advanced concepts were in response to such pressures as (1) the need to decrease excess air in order to increase system efficiency, and

(2) the need to increase combustion staging as a means for
reducing both particulates and NO_x control [61].

Circulating fluidized bed combustors are the primary
advanced combustion techniques that have been developed.
Such fluidized bed designs have been developed by such firms
as the Ahlstrom Company Engineering Division in Karhula,
Finland. Circulating fluid bed units, as depicted schematic-
ally in Fig. 4, have the following features: (1) they can be
operated at high efficiencies through the use of low excess
air levels; (2) they can be operated on blends of dissimilar
fuels such as wood, peat and coal; and (3) they can be built
in large sizes such as the 200 million Btu/hr Ahlstrom power
plant at the Kattua paper mill in Finland shown in Figs. 5
and 6.

Circulating fluidized beds have numerous applications
such as use with high pressure boilers suitable for cogeneration [48]. Further, the ability to burn blends of dissimilar
solid fuels is useful given the periodic shortages of available wood fuel [69].

The Ahlstrom circulating fluidized bed pilot plant went
into operation in 1976. The first commercial unit, a 45,000
lb/hr cogeneration steam boiler retrofit at Pihlava, Finland,
was installed in 1979. Since that time some 600,000 lb/hr of
capacity have been installed at various locations [54]. The
annual capacity growth rate for this system has been 55%.
This unit is sufficiently versatile that it can be used in
applications ranging from pulp mill cogeneration systems to
enhanced oil recovery as shown in Fig. 7.

   2. Gasification Systems. While direct combustion techniques have been advanced, gasification systems have been
developed by such firms as Lamb-Cargate Industries of Vancou-
ver, British Columbia; Omni Fuels of Toronto, Canada; APCO
(Applied Engineering Co.) of Orangeburg, South Carolina;
Combustion Power Company, a Weyerhaeuser subsidiary in Menlo
Park, California; and a host of other firms. Gasification of
wood also has been promoted by such organizations as the
Solar Energy Research Institute.

Gasification involves driving the following equations:

$$\text{Wood} \xrightarrow{\text{heat}} CO_2 + CO + CH_4 + C_xH_y + H_2 + H_2O \qquad (7\text{-}5)$$

$$2C + 1\ 1/2\ O_2 \longrightarrow CO + CO_2 + \text{heat} \qquad (7\text{-}6)$$

$$C + H_2O \longrightarrow CO + H_2 \qquad (7\text{-}7)$$

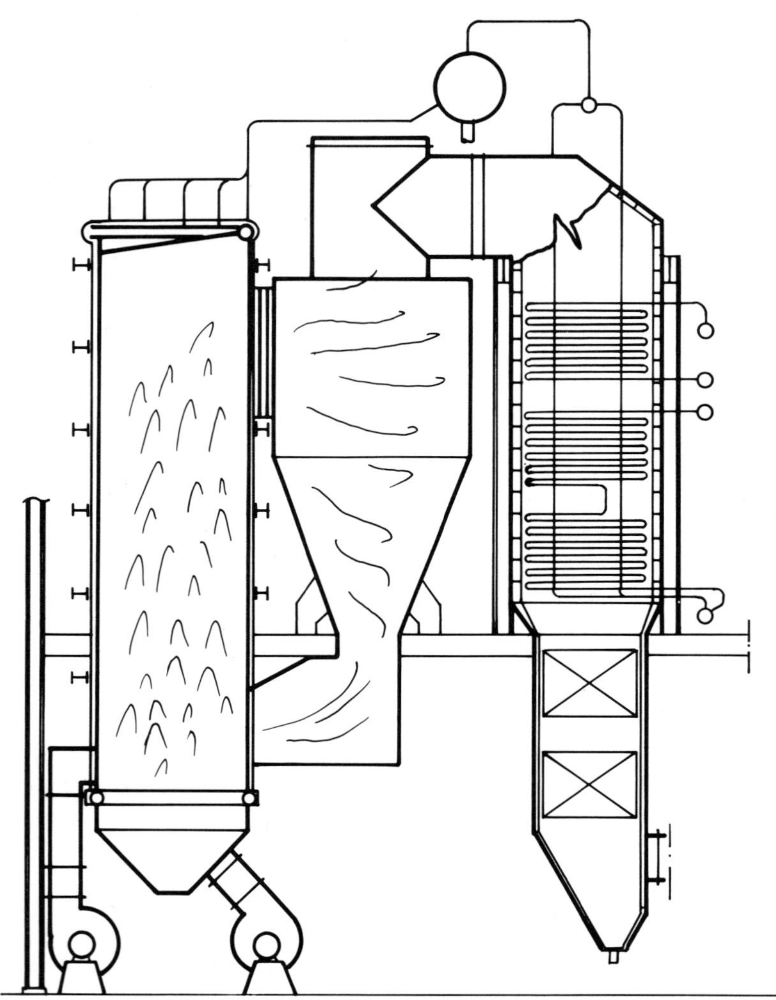

FIGURE 4. Schematic of the Ahlstrom-Pyropower circulating fluidized bed combustor. Source: Leppa [41].

FIGURE 5. Interior view of the Ahlstrom circulating fluidized bed combustor at the Kattua, Finland, pulp mill. (Photo courtesy of Pyropower.)

FIGURE 6. *The Kattua, Finland pulp mill power plant.*
(Photo courtesy of Pyropower.)

$$C + CO_2 \longrightarrow 2CO \qquad\qquad (7\text{-}8)$$

$$C + 2H_2 \longrightarrow CH_4 \qquad\qquad (7\text{-}9)$$

$$CO + H_2O \longrightarrow CO_2 + H_2 \qquad\qquad (7\text{-}10)$$

Gasifiers derive the heat required to drive such endothermic reactions as (7-7) and (7-8) from the heat evolved
from reaction (7-6). Reaction (7-5), which is the basic
pyrolytic reaction, may be endothermic or exothermic depending upon the temperature regime. Gasifiers may use either
air or pure oxygen as the source of O_2 for reaction (7-6).
If air is used to drive reaction (7-6), then the product gas
has a heat content of 110-150 Btu/DSCF. If pure oxygen is
used to drive reaction (7-6), then the product gas contains
300-350 Btu/DSCF. In general oxygen requirements can be calculated as follows [1]: 0.20 - 0.25 lb O_2/lb dry organic
+ 0.31 - 0.36 lb O_2/lb H_2O in feed + 0.32 - 0.37 lb O_2/lb
ash in .feed = total O_2. Higher O_2 requirements exist with
air blown units, and with fluidized bed units, when modest
amounts of the energy released are required for heating inert

FIGURE 7. The Pyropower circulating fluidized bed combustor in Bakersfield, California. This unit is fueled by coal and used for enhanced oil recovery. The coal bunker is the large vessel at the extreme left, the combustor is in the foreground to the right of the bunker and the cyclone is behind the combustor. The baghouse and stack are at the extreme right. (Photo courtesy of Ebasco.)

nitrogen from the air, bed media, or both. The moisture for reactions (7-7) and (7-10) may be supplied by water in the feedstock or by steam injection.

Within the past decade both fixed bed and fluidized bed gasifiers have been designed for industrial applications. Fixed bed units are of both countercurrent and cocurrent design while fluidized bed units typically employ the AFBC concept rather than the circulating fluidized bed approach. Most gasifiers being marketed today use air as the source of oxygen.

Several surveys of wood gasifiers have been conducted recently [13, 32]. These surveys demonstrate that virtually all gasifier concepts available will work when gasifiers are close-coupled to boilers. When electricity generation is contemplated, the typical design is a gasifier coupled to an engine generator. In such cases gas transportability,

high system flexibility, and reliability become critical
criteria.

When the gas produced must be transported tens to hund-
reds of feet before it is combusted, and when flexibility and
reliability are additional parameters, the AFBC gasifier is a
most desirable gasifier. The relative desirability of the
fluidized bed gasifier under such circumstances was demonstrated by Ebasco [13] in its study for the Bonneville Power
Administration, and by Saskatchewan Power Corporation [70],
although the latter organization also experienced success
with fixed bed gasification [68].

The AFBC gasifier can have reasonable feedstock quality
flexibility and reasonable gasification efficiency. Further,
AFBC gasification has the highest rate of fuel throughput per
unit area of grate surface. Additional advantages of AFBC
gasification include: (1) the ability of increased throughput rates to as high as 1500 lb fuel/ft^2 grate/hr by operating
at elevated pressures [49]; (2) the ability to scale up to
capacities in excess of 100×10^6 /Btu/hr without seriously
impacting gasification stoichiometry; and (3) the ability to
recycle all tars, unburned char, and light condensables for
more complete chemical energy recovery [13]. For fluidized
bed designs to be most effective as gasifiers, they should be
based on in-bed rather than over-bed feeding. A comparison
of gasification systems is shown in Table II.

In many respects gasification is a variant on direct combustion, employing the most severe staging possible. Because
gasifiers are extremely staged devices, they are inherently
low producers of NO_x. Gasification may or may not produce
significant quantities of particulates. This advantage of
gasification is best illustrated by examining the second major
gasification system commercialized in recent years: the
Lamb-Cargate close coupled updraft gasifier introduced in
Chapter IV. The Lamb wet cell offers a substantial alternative to fluidized bed systems.

The design and operation of the Lamb-Cargate close coupled gasifier is best visualized in terms of Fig. 8. Fuel at
35-65% moisture content (green basis) is fed by undergrate
ram charger into the gasification chamber. There undergrate
air and steam are fed to the unit in the following ratios:
fuel, 1 lb; air, 1.4-3.8 lbs (depending upon fuel moisture);
and steam, 0.14-0.38 lb. A producer gas with a chemical heat
content of 170-210 Btu/SDCF is produced and passed to the
secondary combustion chamber at temperatures of 1000-1400°F.
This hot producer gas is combusted completely in the secondary combustion chamber. This close coupling of a countercurrent gasifier and the gas combustor improves the system
efficiency by capturing all of the sensible heat in the producer gas and by permitting low excess air combustion.

TABLE II.   Essential Parameters of Typical Wood Gasifiers

Comparison parameter	Gasifier type		
	Fixed bed		Fluidized bed
	Counter-current	Cocurrent	
Feedstock characteristics			
Maximum moisture content	50%	30%	50%
Maximum fines content	12%	12%	100%
Maximum particle size	2"	2"	2"
Throughput			
Maximum size built to date (Btu/h $\times$ 10^6)	25	5	80
Maximum estimated size (Btu/h $\times$ 10^6)	100	N/A	200
Throughput rate (lb fuel/ft^2/h)	20-40	60-100	300+
Gas quality			
Heat content (Btu/SDCF)	150	110	120
Condensables content (%)	<5	<1	<1.5
Typical hot gas thermal efficiencies	75-85%	75-80%	70-80%

Source: Ebasco [13].

Total air to the gasifier and combustor contains on the order of 15% excess air. Consequently exit gases from the combustor are on the order of 2300°F. Exit gas temperatures of 2100°F to 2600°F have been reported. Such gases may be ducted to waste heat boilers for cogeneration of process steam generation, gas fired lumber or veneer kilns, particle

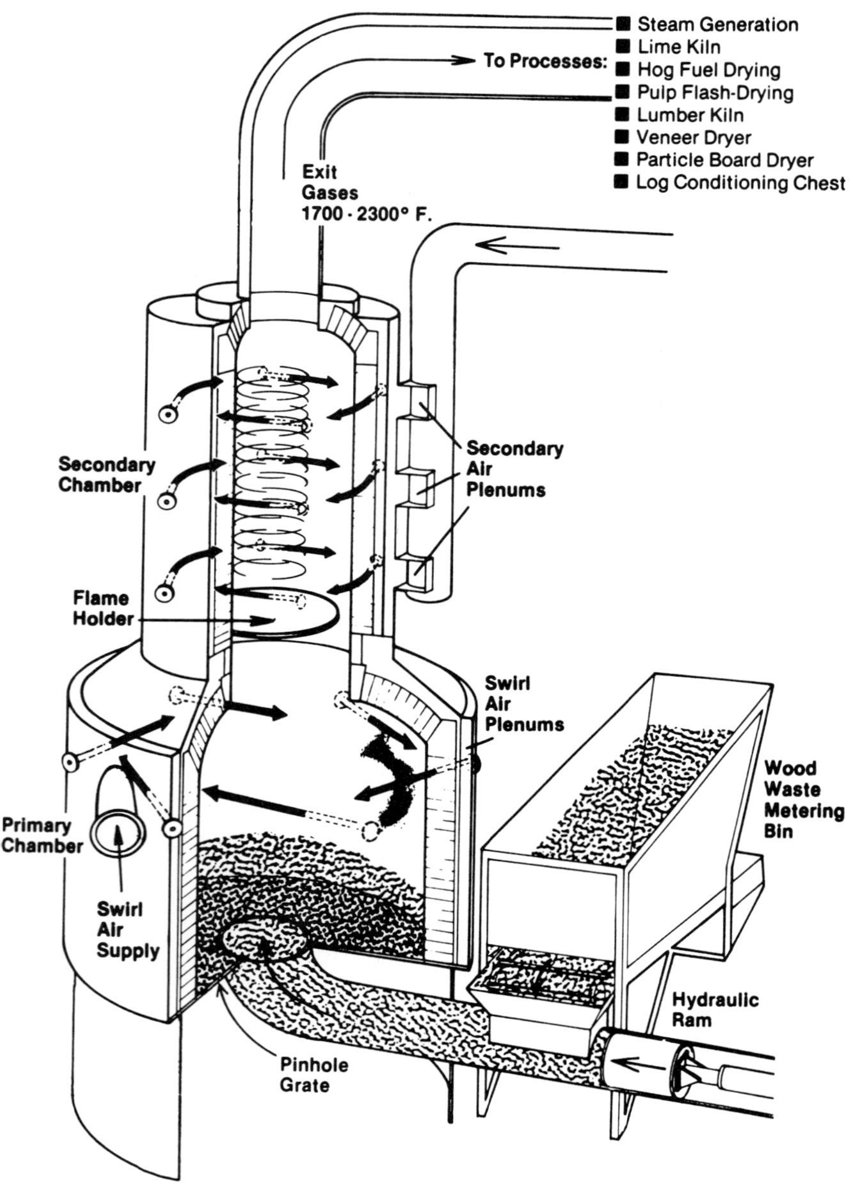

FIGURE 8. Schematic of the Lamb-Cargate close coupled gasifier/combustor or wet cell. Installations of this unit are shown in Chapters IV and V.

dryers for composite product plants, and lime kilns for the
pulp and paper industry (see Chapter V). Thermal efficiencies of such units have equalled 72.5% on 50% moisture fuels,
with final stack temperatures of 300°F. Thermal efficiencies
of well run spreader-stokers burning 50% moisture wood are
<70% based upon 40% excess air and final stack temperatures
of 300°F.

Wet cells can improve the efficiency of wood energy systems. At the same time they provide sufficient combustion
staging and cyclonic action that particulate concentrations
influent to any secondary collector (e.g., baghouse) are
about 0.04 grains/standard dry cubic foot (SDCF) of stack gas.
Emissions of oxides of nitrogen (NO_x) are similarly controlled. The Lamb-Cargate wet cell has become a rapidly
growing wood energy system as discussed in Chapter IV.

Fluidized bed and fixed bed gasification, then, have
emerged as a potentially viable alternative to direct combustion for wood and other biomass fuels. It is particularly
appropriate for such applications as direct heat systems
(e.g., lime kilns in pulp mills, gas fired veneer dryers),
retrofits to exiting gas fired boilers and similar situations.
It may be useful where local air pollution regulations effec-
tively preclude direct combustion of wood. Upon the advent
of commercially proven gas scrubbing systems, wood gasification may become useful in power generation under the right
circumstances.

3. Power Generation Systems. Power generation has
become a major new market for wood fuel with the advent of
PURPA. Some electricity generation systems such as the
Saskatchewan Unit have been based on gasification. Most systems, and most of the breakthroughs,have been in systems
based on direct combustion. Power generation is accomplished
both in cogeneration and stand-alone condensing power plants.
The stand-alone plants are particularly interesting since
they represent a departure from the FPI's tradition of back
pressure turbine cogeneration,

The Burlington, Vermont, Joseph C. McNeill stand-alone
generating station represents the state-of-the-art in wood
fired power generation, and for the following reasons:
(1) it is the most recently constructed unit, having come on
line in late spring, 1984; (2) it is the largest such unit,
having a capacity of 50MW; and (3) it uses potentially the
most expensive fuel, forest residual chips.

The Joseph C. McNeill station accepts forest residual
chips and it burns them in a 500,000 lb/hr Zurn boiler generating 1250 psig/950°F steam. That steam is expanded across
a 50MW Brown Bovari turbine. The system employs two cooling
towers to condense the steam, exhausted at 3.5 to 5.0 in HgA.

The Burlington station burns approximately 80 tons of
fuel per hour. It uses two high pressure and two low pressure shell-and-tube feedwater heaters in addition to the
highly efficient steam generator and turbine in order to promote unit efficiency. The McNeill station also has an extensive tubular air heater to improve boiler efficiency. This
plant has a gross station heat rate of some 13,800 Btu/kWh.
It has a parasitic power load of 3.5MW.

The entire plant, shown previously in Fig. 2, cost $66
million (not including allowance for funds used during construction). That $66 million includes all studies, permit-
ting, land, and owner related costs. The plant is operated
by five 6-man work crews plus 10 maintenance, supervisory,
and related personnel. It pays about $2.00/Btu $\times$ 10^6 for
fuel. Its 1984 annual expenses, therefore, can be estimated
as follows (at 6500 hrs/yr): fixed operation and maintenance,
$2.65 $\times$ 10^6/yr; and variable operation and maintenance,
$8.97 $\times$ 10^6/yr.

Power production costs can then be estimated assuming an
investor owned utility, a 30-year project life, a nominal
after-tax cost of capital of 12%, and an estimated rate of
inflation of 5%. The levelized costs are shown in Table III.
The Burlington costs are lower because the facility is owned
by a municipality with a low cost of money and no tax
obligations.

TABLE III. *The Levelized Cost of Power From a 50 MW
Investor-Owned Utility Power Plant (Cost Basis:
Burlington, Vt. Capital and O & M Costs)*

	Dollar base	
Cost	Nominal	Real
Capital charges	9.9×10^6	6.6×10^6
Fixed operating and maintenance	4.22×10^6	2.65×10^6
Variable operating and maintenance	14.3×10^6	8.97×10^6
Total	28.4×10^6	18.2×10^6
kWh produced (net)	302.25×10^6	302.25×10^6
Levelized cost (¢/kWh)	9.4	6.0

Wood fired power plants come in all sizes, from 5MW to
50MW. The typical costs for such units not including AFUDC,
client (owner) charges, land, permitting, and related costs,
are shown in Table IV. Within these size ranges a general
exponential scale factor of 0.8 is appropriate. Below 5MW
the scale factor declines to the range of 0.6-0.65.

Wood fired power plants are a growing phenomenon. In
addition to the plants existing today, wood fired power plants
have been proposed for and are under evaluation in such com-
munities as Tallahassee, Florida, Delano and Stockton, California, White City, Oregon, and Seattle, Washington. The
situation today is vastly different from the years when the
Eugene, Oregon, Water and Electric Board (EWEB) was the only
utility with wood fired capacity.

Power generation by cogeneration and condensing power is
a vast market for wood. Wood fired power plants have become
economically acceptable in some portions of the country because they can be built in reasonable sizes, and in short
time frames (e.g., 2-5 years). Like advanced combustion and
gasification systems, they provide a basis for expanding the
use of low value wood materials.

B.   *The State-of-the-Art in Chemicals and Related
     Products Manufacture*

With the recent activity in energy, it is not uncommon
for analysts to overlook wood as a chemical feedstock. Wood,
however, supplies the economy with a substantial quantity of
specialty chemical and related products, including foodstuffs
for humans and cattle. Like the production of cogenerated
electricity, the production of chemicals currently is largely

*TABLE IV. Capital Costs for Wood Fired Power Plants
   (Basis: Overnight 1984 $)*

Unit size	Capital cost (1984)
5 MW	9,000,000
25 MW	33,000,000
50 MW	58,000,000

*Sources: 25 MW unit, Ebasco [15]; 50 MW unit, analysis
of Burlington, Vt. values.*

a function of pulp mill activity. Chemicals produced are
direct products of the pulping process (e.g., dissolving pulp
products) and byproducts such as turpentine and tall oil,
lignosulfonates, ethanol, and torula yeast.

    1. Dissolving Pulp Products. Alpha-cellulose pulp pro-
duction is a highly significant sector of the wood chemicals
industry. Dissolving pulp is used to manufacture textile
fibers, chemicals, plastics, very high grade papers, and
other products [45]. Currently the U.S. has some 1.4×10^6
ton/yr of dissolving pulp capacity [51]. Production is in a
long term decline, but it currently exceeds 1×10^6 ton/yr
(estimated from [45]).
    Dissolving pulp is produced either by sulfite pulping or
kraft pulping. If kraft pulping is employed, the pulp chips
undergo an acid prehydrolysis step. Severe bleaching also is
involved in α-cellulose production as the product must be
free of lignin, hemicelluloses, and ash. Because of the
severity of pulping and bleaching conditions, dissolving pulp
yields are in the 28-38% region [62].
    Products commonly associated with dissolving pulp and
viscose include rayon, cellophane, cellulose acetate, cellulose nitrate, cellulose esters for thermoplastics, carboxy-
methyl cellulose for paints, and several others. Of these,
rayon is the dominant product, as over 200×10^3 ton/yr are
produced. Rayon is successful because it maximizes user com-
fort and can be blended with easy-care petrochemical based
synthetic fibers [34]. The advantages of rayon make it a
specialty product, consistent with all wood chemicals. In
total, fibers and films produced from dissolving pulps comprise less than 10% of the total synthetic fibers industry
[62]. Dissolving pulp and associated products contributes
750×10^6 - $\$1 \times 10^9$ to FPI revenues.

    2. Extractive Based Chemicals. Wood is composed not
only of cellulose, the hemicelluloses, and lignin, but also
of up to 5% extractives (e.g., α-pinene, β-pinene). These
extractives provided the basis for the original Naval Stores
industry. Trees were slashed or "wounded" and tapped for
gum, a composite of numerous extractives. The gum, once
gathered, could be steam distilled into turpentine and rosin.
    The Naval Stores industry concentrated in the Southeast,
emphasizing production from two Southern Pine species: slash
pine and longleaf pine. In an effort to reduce raw material
costs, the industry began focusing production efforts on
stumps from harvested pine trees. When the kraft pulp industry became a dominant economic force in the southeast, it
became the dominant supplier of such extractive based
chemicals.

Even today all three production pathways exist for manufacturing Naval Stores, as Fig. 9 shows. Kraft pulping began
supplying over 50% of all Naval Stores by 1970 [8]. Rosin
from tree wounding (gum rosin) now comprises <2% of the total
rosin production. Gum turpentine comprises slightly over 1%
of all turpentine produced in the U.S. [66].

The growth of pulp mills as a supplier of extractive based
chemicals is largely a function of the growth of sulfate
pulping in the south. Currently softwood kraft pulping capacity in the south approaches 25×10^6 ton/yr, or 76% of
U.S. softwood kraft pulping capacity [57]. The only limiting
factor on extractive chemicals production from kraft mills
appears to be the growth of continuous digester capacity (see
Chapter IV), since continuous digesters are less efficient
producers of Naval Stores than batch digesters [57].

The extractives based industry is, at best, stagnant.
Today some 700×10^6 lbs of Naval Stores are produced
annually, which is consistent with production levels for the
past decade [66]. Production has not exceeded 1×10^9 lbs
since 1966 [8]. While the industry is stagnant in volume of
product, however, the value of its output was 300×10^6 in
1980 [66], and approached 400×10^6 in 1984. Of this total,
about 20% is turpentine and 80% is rosin [66]. These revenues are based upon the versatility of extractive based chemicals as shown in Table V. It is useful to note the uses of
some extractive based products. Pine oil, for example, is
used as a chemical in froth flotation concentration of min-
erals (e.g., calcium carbonate), in disinfectants and clean-
ers, in deordorants, and in a host of other applications.
The α-pinene from sulfate turpentine is used to make synthetic pine oil and camphor.

While wood extractive based chemicals are not obvious to
the casual observer, they are a significant FPI business.
They supply functions not easily achieved by chemicals syn-
thesized from oil.

3. Lignin Based Chemicals. Lignin chemicals also are
based on pulp mill byproducts—spent pulping liquors. Lignin
chemicals offer dramatic growth for the FPI. In 1960 about
2×10^6 worth of lignin chemicals were produced, and by 1980
that value had risen to 180×10^6 [43]. Lignin chemicals
have experienced a compound annual growth rate of 25%/yr (see
Fig. 10). This growth of lignin chemicals has to be considered phenomenal in an industry largely characterized by slow
growth if not stagnation.

Lignin chemicals produced from kraft and sulfite liquors
include animal feed binders, oil well drilling mud additives,
concrete additives, dyestuff dispersants, agricultural chemicals, tanning agents, food flavoring agents, solvents, and

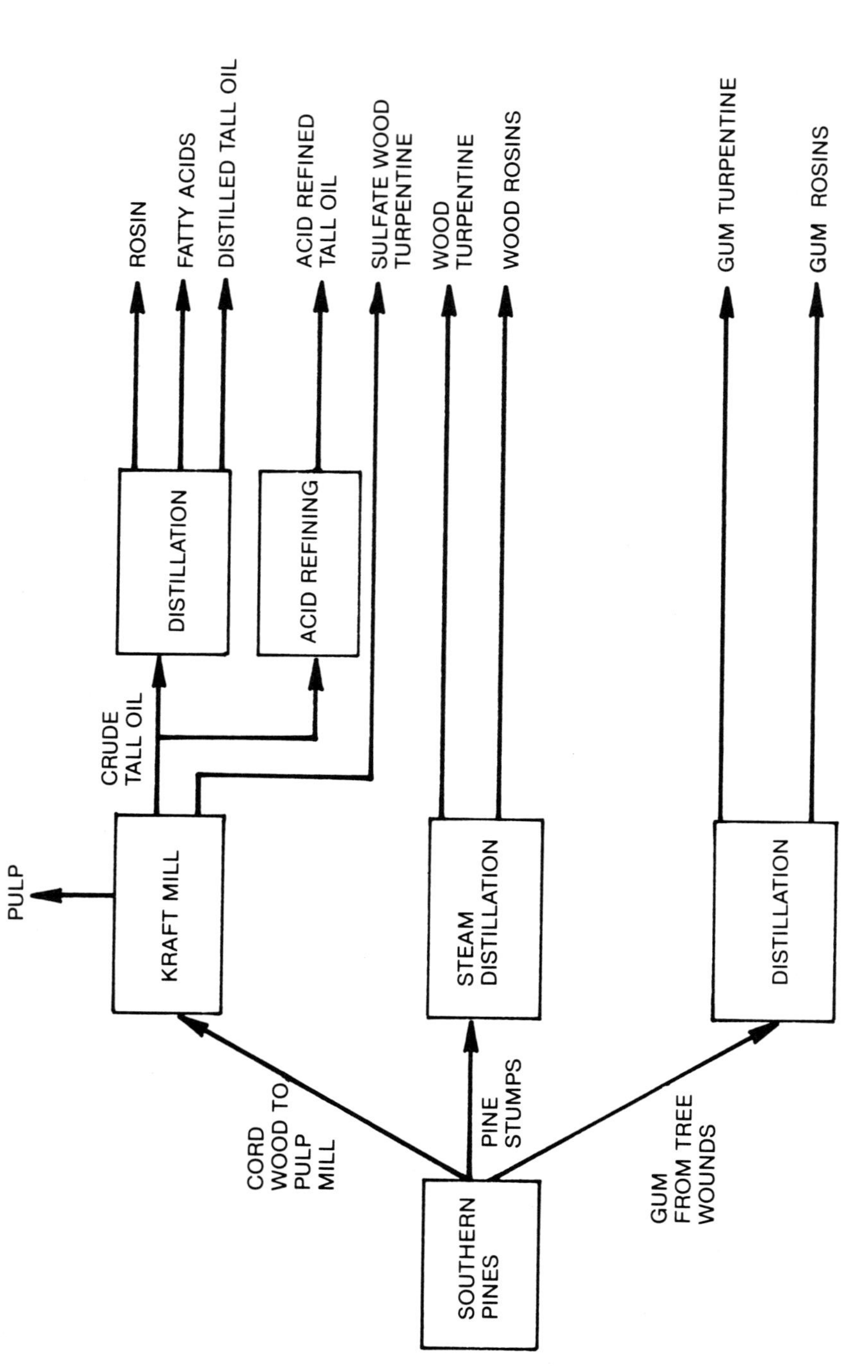

FIGURE 9. Pathways for the production of Naval Stores.

TABLE   V.   Uses of Extractive Based Wood Chemicals

Naval Store	Use	Percent
Rosin	Rubber and chemicals manufacture	35.2
	Paper sizing	33.5
	Ester gums and synthetic resins	22.7
	Paints and lacquers	2.2
	Other, miscellaneous	6.7
Turpentine	Synthesis of pine oil	48.0
	Manufacture of turpene resins	16.0
	Turpene insecticides	15.0
	Flavors and fragrances	10.0
	Other, miscellaneous	11.0

Source: Bratt [8].

medicinal compounds [8, 43]. Table VI identifies some of
these products. Among the more recognized products made from
lignin are dimethyl sulfide, dimethyl sulfoxide (DMSO), lignosulfonates, and vanillan.

Crown Zellerbach as the capacity to produce 20×10^6 lb/
yr of dimethyl sulfide at its Bogalusa, Louisiana, kraft mill.
The original pilot plant for DMS production was built in
1956-1957, and the first 10×10^6 lb/yr of capacity was installed at Bogalusa in 1960 [26]. The DMS plant of Crown
Zellerbach also has facilities to oxidize that product to
DMSO. The oxidation is achieved by reacting DMS with NO_2
[26]. That capacity to produce DMSO is particularly important. DMSO has medicinal properties although it was developed originally as a highly useful industrial solvent.

Westvaco produces lignosulfonates by acidification of
kraft black liquor and then sulfonation of that intermediate,
sulfate lignin. Other firms obtain lignosulfonates directly
from sulfite liquor [8]. Vanillan is produced from spent
sulfite liquor by such firms as Monsanto Chemical, Seattle,
Washington [26].

The growth of lignin chemicals in the face of energy
costs for wood fuels (including spent pulping liquor) of
$2/Btu $\times 10^6$ illustrates the interrelationships among energy,

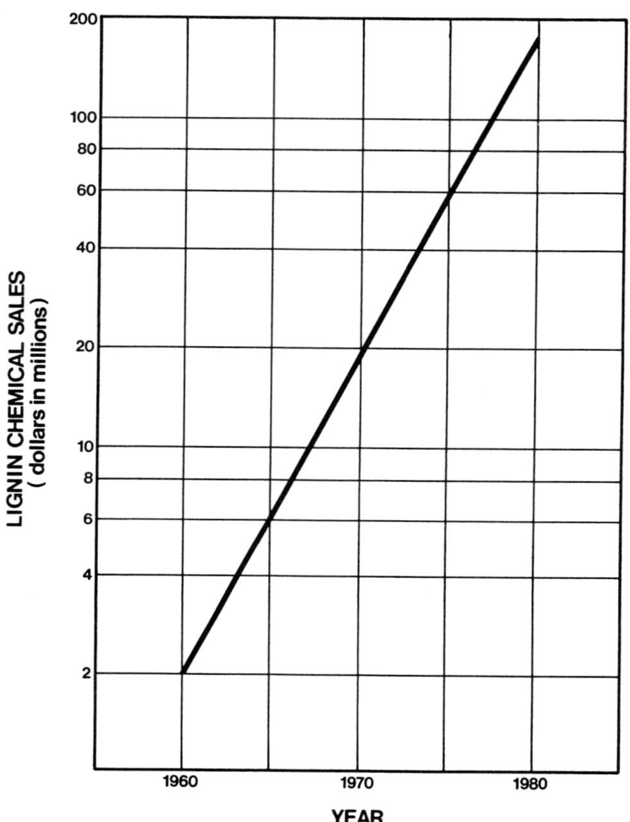

FIGURE 10. The growth of lignin chemicals.
Source: Lin [43].

chemicals, and related products. The feedstock for lignin
chemicals now has a significant opportunity cost that must be
factored into its economics. This phenomenon is discussed
completely by Frederick [21].

4. *Other Byproduct Chemicals*. Pulp mills produce a host
of miscellaneous byproduct chemicals from the sugars contained in spent pulping liquor. Some examples are discussed
below.

About 5×10^6 gal/yr of ethanol is produced by fermentation of the sugars in sulfite liquor. This EtOH production
occurs at the Georgia Pacific mill in Bellingham, Washington.
Monarch Paper of Rhinelander, Wisconsin, and Boise Cascade at
Salem, Oregon, produce torula food yeast. This foodstuff

TABLE VI. Current Application Areas of Lignin Based Chemicals

Current application areas	Performance property
1. Carbon black and pigment dispersion	
2. Cement and concrete	
3. Dyestuff formulations	
4. Expander in lead and acid batteries	Dispersing
5. Gypsum wallboard	
6. Wettable & flowable pesticides	
7. Oil well drilling	
8. Boiler and cooling water treatments	
9. Corrosion inhibition	
10. Industrial cleaner	Complexing/ Dispersing
11. Micronutrients	
12. Protein precipitation	
13. Adhesives for board and veneer	
14. Animal and poultry feed pellets	
15. Ceramic and refractories	
16. Foundry sands	Binding
17. Ore and coal briquettes	
18. Soil conditioning	
19. Asphalt emulsion	Emulsion stabilizing
20. Wax stabilizer	
21. Fire log	
22. Fire fighting foam	
23. Molasses conditioning	

Source: Lin [43].

contains 50% protein along with B-complex vitamins, and is
available in many health foodstores. About 20×10^6 lb/yr may
be produced. Sunoco Products produces acetic acid from spent
pulping liquor at Hartsville, South Carolina. The pulping
liquor is acidified. Extraction with methyl-ethylketone (MEK)
follows acidification [8].

Of the chemicals identified above, the ethanol from
Bellingham is perhaps most interesting. For a variety of
legal/regulatory reasons, it cannot be sold as a foodstuff.
It is sold for industrial applications. When the gasohol
movement erupted in the mid 1970s, the Bellingham mill was
the prime source of alcohol. It held that position until
agricultural crop based capacity came on line.

5. Summary of Chemicals Manufacture. The production of
chemicals from wood is a diverse if not diffuse business. At
the same time, there are certain economic constants worth
noting. The wood chemicals industry is highly related to,
and dependent upon, the wood pulping industry. This holds
true regardless of type of product being manufactured:
α-cellulose, extractive chemical, lignin chemical, or miscellaneous products (e.g., Torula yeast). Chemicals production
demonstrate again just how central the pulp mill is to the
forest products industry and the integrated mill.

The second critical point concerning wood chemicals is
that they do not compete in the bulk chemical market. Wood
currently does not compete as a feedstock for such commodities as ethylene, benzene, ammonia, and other similar chemicals now derived from petroleum and natural gas. The wood
chemical products such as ethanol from Bellingham, rayon, and
DMSO from Bogalusa illustrate the principles of successful
wood chemicals and related products: (1) they are produced
in less than bulk quantities; (2) they serve carefully nur-
tured specialized markets; and (3) they are produced from low
value raw materials (e.g., as byproducts).

III. CONCEPTS FOR ADVANCED PRODUCTION OF ENERGY, CHEMICALS AND RELATED PRODUCTS

Numerous innovations have been proposed for producing
energy, chemicals, and such related products as cattle and
human food from wood. All of these innovations are designed
either to expand the role of wood within existing markets
(e.g., electricity generation) or to open up new markets
(e.g., dietetic foods). In many cases these proposed innovations are as related to one area (e.g., energy) as another
area (e.g., chemicals). Such "cross over" cases include

oxygen blown gasifiers and other systems for producing low
molecular weight alcohols. This chapter presents some representative wood-to-energy/chemical concepts. For convenience,
it is organized as follows: (1) electricity technologies,
(2) liquid fuels and chemicals technologies, and (3) foodstuff technologies.

A. *Electricity Generating Technologies*

Electricity represents a growth business for the FPI,
both through the sale of fuel to utilities such as Washington
Water Power, and through the sale of cogenerated power. The
sale of electricity holds more potential for the FPI because
the marginal value of wood fuel is $2/Btu $\times 10^6$, while elec-
tricity valued at 5¢/kWh is worth $14.65/Btu $\times 10^6$. Elec-
tricity generation is significant to the FPI for internal
reasons as well as PURPA markets. The FPI is altering its
own electricity/process heat consumption balance in favor of
kilowatts. The growth of TMP pulping is a most visible mani-
festation of this shift towards more electricity and less
process heat in the FPI.

The growth in importance of electricity generation for
the FPI has led to numerous alternative cogeneration concepts. These cogeneration concepts address some of the limitations associated with the steam turbine topping cycle
(Rankine cycle) that now dominates FPI cogeneration activity.
The Rankine cycle is the most thermodynamically efficient
cogeneration system for wood fuel [63]. However it has the
lowest electricity/process heat ratio of the major cogeneration cycles, the highest capital cost of all cogeneration
systems, either on a total cost or incremental cost basis
unless very small systems (e.g., <1MW) are contemplated [13].
One problem is that the Rankine cycle system must treat all
water for use in the high pressure boiler, despite the fact
that up to 40% of the steam sent to process may not return as
condensate [36], creating high capital and operating expenses
in this critical area.

Four alternative cogeneration systems have been proposed
for the FPI: (1) the "downheat" or air Brayton cycle, where
compressed air is heated in a wood combustor and used as the
working fluid in a gas turbine; (2) the wood fired combustion
turbine, where clean dried sawdust is fired directly into a
gas turbine instead of natural gas or oil; (3) the wood gasifier based combustion turbine where wood is gasified, the gas
is cleaned and cooled for higher chemical energy density, and
the gas is then burned in a combustion turbine; and (4) the
wood gasifier based reciprocating engine (e.g., Diesel cycle,

Otto cycle) where either a spark ignition or diesel engine is
substituted for a combustion turbine. (See Fig. 11.)

The first three options are of immediate interest and use
to the FPI. The fourth option is largely ineffective for the
FPI because it produces process heat as hot water. It could
be used in conjunction with the Rankine cycle, where the gas
conditioning system and cooling water jacket could serve as
boiler feedwater heaters. Because the first three options
are of more immediate interest, certain salient characteristics are summarized in Table VII. It is clear from Table VII
that the air turbine or downheat Brayton cycle system represents the best compromise in terms of power/heat and fuel
cost chargeable to power. Thus the air turbine system is
analyzed in detail in Section IV. It is the wood fired Brayton cycle system that is closest to commercial deployment.
Further it captures the benefits of advanced combustors such
as the wet cell and the circulating fluidized bed.

The pulverized wood fired gas turbine also offers
intriguing possibilities. It provides for a simple yet potentially efficient system. Consequently, it is being pursued
vigorously by the U.S. Department of Energy (DOE). In principle this system involves burning finely divided wood parti-
cles in suspension, in a stream of compressed air. The products of combustion are passed through a hot cyclone and then
to the combustion turbine [31].

A Garrett 500 horsepower turbine has been run on clean,
dried, pulverized sawdust with promising results [44]. The
testing period to date has been short; however, no erosion of
nozzles or blades was apparent after 200 hours of operation.
The research is now being scaled up, and a 3MW Allison turbine is being tested in this system [31]. The results of
tests in excess of 1000 and 10,000 hours, and at the larger
scales, now are necessary to prove that this concept is tech-
nically feasible. The technical promise of this concept is
not totally unexpected, however. Currently conventionally
fired gas turbines are cleaned with walnut shells.

B. Liquid Fuels and Chemicals

While wood has been proposed as a fuel to be used for
electricity, it also has been proposed as a feedstock for the
production of liquid fuels and chemicals. During the 1970s
and early 1980s, considerable effort was spent to develop
biochemical and thermochemical conversion systems for the
manufacture of such products.

Biochemical processes have been designed to produce
ethanol or methane from wood, with ethanol receiving the
dominant attention. Thermochemical processes have been

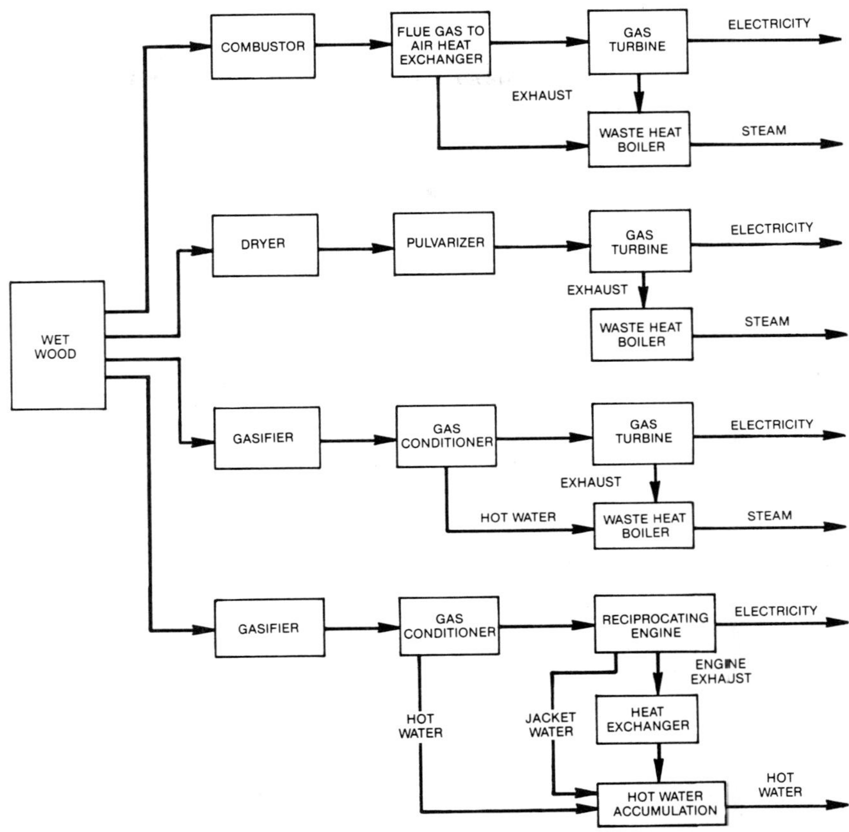

FIGURE 11. *Alternative wood fired cogeneration systems.*

TABLE VII. Selected Parameters of Alternative Cogeneration Cycles

	Cogeneration Cycle			
Parameter	Backpressure steam turbine	Air turbine (Downheat Brayton cycle)	Direct fired combustion turbine	Gasifier based combustion turbine
Electricity/heat ratio (kWh/ 50 psig steam)	.06	.09	.17	.16
Fuel cost chargeable to power (Btu/kWh)	4800–5200	7500–8700	10,500–15,200[a]	11,300–14,100
Development remaining	None	Proof of flue gas to air heat exchanger	Testing turbine blades for erosion	Development of efficient gas conditioning

[a] Includes energy required for drying.

Source: Tillman, Schnorr, and Sale [63].

designed to produce a wide variety of products from methanol
to aromatic chemicals.

Biochemical pathways have focused on enzymatic and acid
hydrolysis of wood, followed by fermentation processes yielding ethanol. Excellent reviews of this technology have been
published by Emert [16], and previously by Gaden *et al.* [23].
In general it is apparent that delignified substrates such as
municipal wastes or some agricultural materials are superior
to wood in the production of ethanol (or methane).

Thermochemical processes have more versatility and diversity than biochemical pathways, as is highlighted in Fig. 12.
Two basic approaches exist: (1) gasification followed by indirect liquefaction or other C_1 chemistry processes; and
(2) pyrolysis (deoxygenation), possibly followed by hydrogenation. Of these pryolysis (without hydrogenation) has been
practiced commercially. Pyrolysis reactors included the
Stafford-Badger retorts employed by the Ford Motor Co. at
Iron Mountain, Michigan, from 1924 through World War II. Ford
produced charcoal, fuel gas, ethyl acetate, and other chemicals [59]. Despite historical precedent, indirect liquefaction appears to be more popular today. The versatility of
indirect liquefaction in process and product, and the ability
to achieve technology transfer from coal to wood are reasons
for its attractiveness (see Probstein and Hicks [52]).

Indirect liquefaction depends upon the production of synthesis gas, and several wood based systems have been proposed.
Katzen Associates [39] proposed the generation of low Btu
(producer) gas with the Moore-Canada fixed bed updraft gasifier. The nitrogen in the product gas was removed by cryo-
genic separation [56]. Mitre Corporation [6] proposed use of
the fixed bed oxygen blown Purox[tm] updraft gasifier made by
Union Carbide Corporation. More recently, the Solar Energy
Research Institute has developed a fixed bed downdraft oxygen
blown gasifier as shown in Fig. 13. This gasifier produces a
synthesis gas with a high CO/CO_2 ratio, a low CH_4 content,
and less than 1% tar [55]. It was designed for use in
methanol synthesis. Additionally Inco Metals developed a
design for an oxygen blown fluidized bed biomass gasifier [9].
Battelle-Columbus laboratories also has developed a fluidized
bed system for synthesis gas production, and this gasifier
does not require a pure oxygen supply [18, 50].

The synthesis gas produced from various wood gasification
systems can be used to produce methanol according to any of
the following reactions [56, 71]:

$$CO + 2H_2 \longrightarrow CH_3OH \qquad\qquad (7\text{-}11)$$

$$CO_2 + 3H_2 \longrightarrow CH_3OH + H_2O \qquad\qquad (7\text{-}12)$$

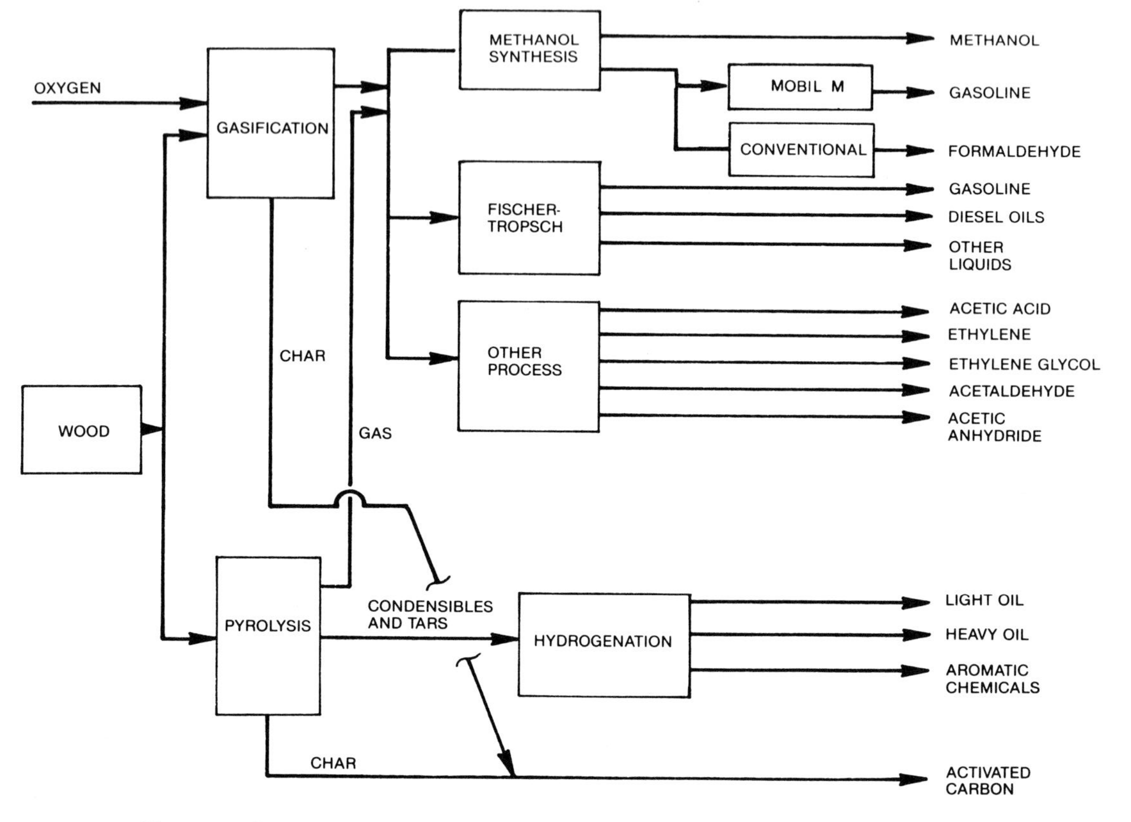

FIGURE 12. Pathways for producing of liquid fuels and chemicals from wood.

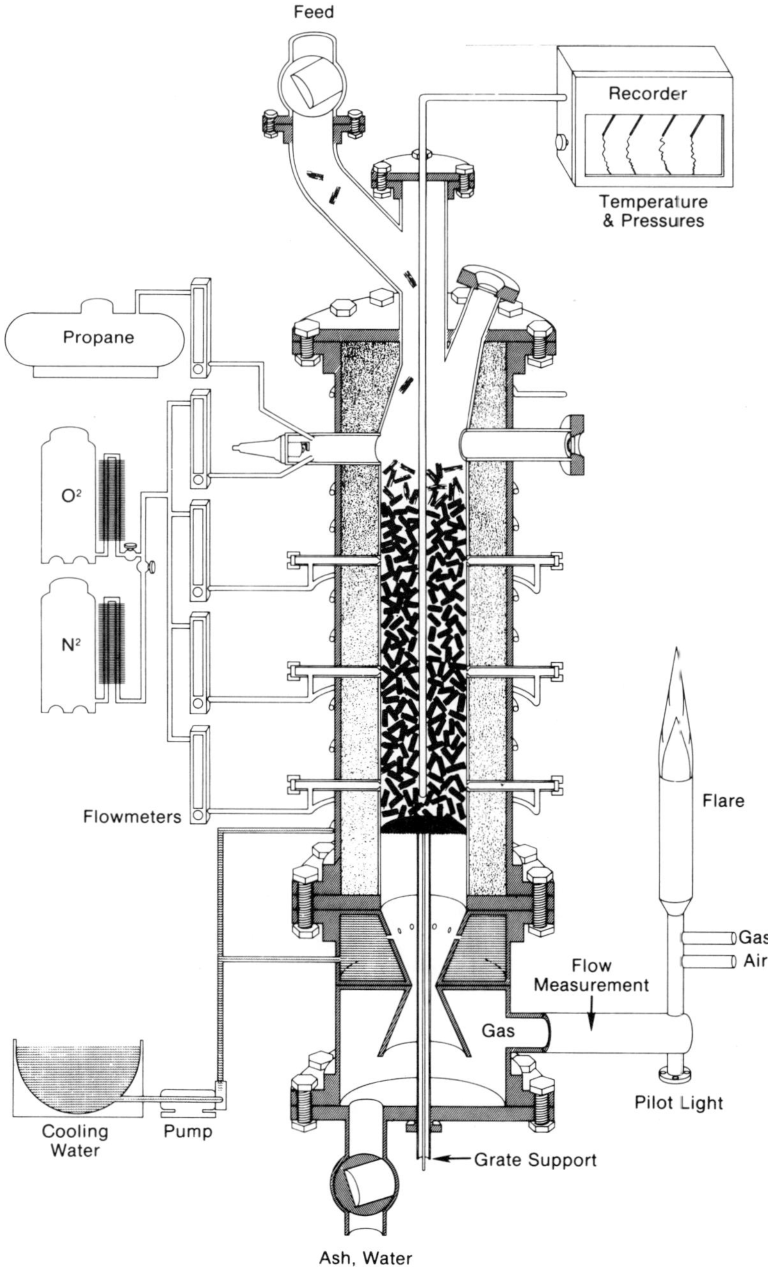

FIGURE 13. The SERI oxygen blown biomass gasifier.

$$3CH_4 + CO_2 + 2H_2O \longrightarrow 8H_2 + 4CO \longrightarrow 4CH_3OH \qquad (7\text{-}13)$$

The methanol produced can be used directly or it can be employed as an intermediate in the production of formaldehyde according to the following reactions [71]:

$$CH_3OH + 1/2\ O_2 \longrightarrow CH_2O + H_2O \qquad (7\text{-}14)$$

$$CH_3OH \longrightarrow CH_2O + H_2 \qquad (7\text{-}15)$$

The production of formaldehyde consumes 90% of the methanol now produced [20]. Methanol can be used to produce a variety of products beside formaldehyde, including gasoline by the Mobil M process, methyl halides, methylamines, acetic acid, methyl tertiary butyl ether, styrene, ethylene, ethylene glycol, and single cell protein [20]. If formaldehyde is produced, it can be converted into the class of engineered plastics known as polyformaldehydes [7].

This production of methanol, gasoline, and related chemical products highlights the transferability of technology from coal conversion to forest products. This transferability includes application of Fischer-Tropsch (F-T) chemistry to synthesis gas made from wood. F-T chemistry has been commercialized and is currently being used at the Sasol facilities in South Africa (see Fig. 14). The current processes are direct descendants of the coal conversion processes developed during the late 1920s and 1930s in Germany (see Irgolic *et al.* [37], Donath and Hoering [10]). The transferability of this technology to wood is eased by the absence of sulfur in wood, hence the elimination of the need to require sulfur resistant catalysts.

In addition to the production of methanol, synthesis gas can be used to produce other chemicals including the following [3]:

$$\text{Acetic acid} \qquad 2H_2 + 2CO \longrightarrow CH_3C(O)OH \qquad (7\text{-}16)$$

$$\text{Ethylene glycol} \qquad 3H_2 + 2CO \longrightarrow HOC_2H_4OH \qquad (7\text{-}17)$$

$$\text{Ethylene} \qquad 4H_2 + 2CO \longrightarrow C_2H_4 + 2H_2O \qquad (7\text{-}18)$$

$$\text{Acetaldehyde} \qquad 3H_2 + 2CO \longrightarrow CH_3CHO + H_2O \qquad (7\text{-}19)$$

Synthesis gas, then, is a highly versatile chemical feedstock that can be used in a myriad of ways (see Haggin and Aquilo *et al.* [29, 3], for a complete review).

Liquid fuels and chemicals from wood, produced either by biochemical or thermochemical means, are not competitive today. Further, they have little appeal in the immediate

FIGURE 14. The Fischer-Tropsch reactors at the Sasol facility in South Africa. Indirect liquefaction can be used for wood as well as coal.

future due to stability in the world price of oil. Rising
real prices of world oil as forecast for the end of this
decade by the U.S. Department of Energy [67] may make such
processes far more attractive, for the long term.

C. *Extractive Based Chemicals*

Production techniques have been developed not only to
improve production of chemicals such as methanol, but also
specialty chemicals such as turpentine and rosin. Extractive
based chemical production can be enhanced by wounding any
southern pine and injecting solutions containing 2-4% paraquat. The result of such injections is lightwood, extractive
soaked wood.

The discovery of lightwood production by paraquat injection occurred in 1965 at the Olustec, Florida, laboratory of
the U.S. Forest Service. Since that time the Forest Service
and private corporations such as Union Camp have determined
that turpentine yield per cord of wood increases from 1.2
gal/cord to about 4 gal/cord within 24 months of paraquat
treatment, that α-pinene yield increases from 0.75 to 2.26
gal/cord, and that β-pinene yield increases from 0.23 to 0.81
gal/cord [5]. Fatty acid yield appears substantially un-
affected by this process [5].

Paraquat injection does not apparently affect methods
appropriate for the pulpwood harvest or yields obtained from
kraft pulping [58]. Consequently, this method of Naval
Stores production is available and may well grow rapidly.
Alternatively Naval Stores may be produced by harvesting the
tap roots of southern pines [40].

D. *Foodstuffs From Wood*

The final category of wood products proposed is foodstuffs for both cattle and humans. Such foodstuffs have
market appeal because they can take the pressure off agricultural products. Wood can contribute to food supply in a
country facing the steady loss of topsoil and in a world with
population problems. Whether or not such processes are economically attractive, however, remains at issue.

1. Cattle Food Production. Currently sawdust is sold to
feedlots as a cattle food supplement [12]. Cattle can consume up to 5% to 15% of this material as roughage [30]. Wood
is virtually indigestible, even to ruminant animals, however.
Consequently no food value is imparted with this approach.

By 1920 experiments were underway to treat wood with 1.8%
sulfuric acid in order to improve digestibility [30]. More
recently it has been discovered that the digestibility of
woods—particularly aspen—increases dramatically by treatment with ammonia, sulfur dioxide, sodium hydroxide, or white-
rot fungi. Alternatively the wood can be treated with elec-
tron radiation or by vibratory ball milling [4]. Ammonia
treatment can be in liquid or gaseous form. It increases the
protein content of the wood from 0.5% to 9% [4].

Steam extraction has been proposed as an alternative to
chemical treatment for producing animal foodstuffs [53].
Steaming followed by water extraction raised aspen digestibility from 6 to 65%, oak digestibility from 7 to 74%, and
birch digestibility from 5 to 71% [53]. The Stake and IOTEC
processes are alternative concepts for producing cattle food
from wood. These are steam explosion based approaches [62].

Because food production is of critical importance, cattle
food manufacture is the second innovation analyzed in this
chapter. It demonstrates techniques for evaluating high risk
(new product and new process) innovations.

2. Human Food. Of final interest is the production of
human food and, in particular, dietetic food supplements from
wood pulp. ITT Rayonier has developed a new product called
"Microfibrillated cellulose" (see Herrick *et al.*[33], Turbak
et al. [65]). This product is made by beating cellulose wood
pulp in order to expand its surface area and open up the sub-
structural microfibrils. Typical pulps used in the experiments are high cellulose content sulfite pulps, although
there is no reason to suspect that highly delignified organisolv pulps would not be adequate for the product.

Microfibrillated cellulose (MFC) contains 2–4% solids. It
can be used in a wide variety of products. Representative MFC
products are listed in Table VIII. Such food products range
from staples and meat supplements to desserts. Additional
uses involve manufacture of paints, cosmetics, paper, tex-
tiles, and medicinal goods [65]. ITT estimates that the
costs to produce 2% MFC are 1.5¢/lb ($30/ton) including capi-
tal charges, all raw material (e.g., sulfite pulp), and other
operating costs.

*E. Summary of Opportunities in Energy, Chemicals,
    and Related Products*

There are numerous new technologies available in such
areas as cogeneration, liquid fuels and chemicals synthesis,
Naval Stores production, cattle food manufacture, and even
human food manufacture. Such opportunities may enhance the

TABLE VIII. A Representative List of Foods Made with Microfibrillated Cellulose

---

Fruit creams

Ice cream

Gelatin desserts

Cream soups

Gravy

Onion dip

Pudding

Salad dressing

Hamburger supplement

Cake frosting

Source: Turbak et al. [65].

quietly large businesses of energy, dissolving pulp, Naval Stores, lignin based chemicals, and miscellaneous chemicals which now comprise a $6.5 billion dollar segment of the FPI.

The greatest growth opportunities for the FPI may be in these areas of energy, chemicals, and foodstuffs. Consequently the final innovations analyzed in this text are a novel cogeneration system and a cattle food production method.

IV.   DETAILED NEW PROCESS ANALYSES

The downheat or air turbine Brayton cycle and the chemical based production of cattle food represent vastly different levels of commercial proximity, technical and business risk, and industry position. Such differences are reflected in the analyses below.

A.   *The Downheat Brayton Cycle or Air Turbine System*

The air turbine system has been proposed in numerous publications (see, for example, Fryling [22]; Tillman, Rossi, and Kitto [61]; Mills and Peltier [46]) as a means for using solid fuels with combustion turbines. It has distinct cost and product advantages as outlined previously. With the advent of PURPA, and the marketability of cogenerated electricity, it has become increasingly important.

In principle, the downheat Brayton cycle employs a conventional wood combustor, and transfers the heat from the products of combustion to compressed air. This heated compressed air then is used to drive the combustion turbine. Exhaust from both the flue gas-to-air heat exchange system and the combustion turbine can be ducted to one or more waste heat boilers in order to raise steam for additional power generation in a conventional (or induction) back pressure turbine.

1. Recent Developments in Air Turbine Systems. The air turbine system can use either wood or coal as a fuel. Although much of the research was conducted assuming wood as fuel [27, 28], considerable effort has been made to use coal as well [19, 47]. In both the cases of wood and coal, the critical system components are the combustor and the heat exchanger.

The advanced wood combustion technologies have contributed significantly to the desirability of using wood fuels in air turbine systems. The advanced gasification/combustion system shown in Fig. 8 has been proposed as one desirable combustor for the downheat system as it can raise products of combustion in the 2100–2600°F temperature region. Such a thermal head improves the heat transfer efficiency within the system. This approach is shown by Mills and Peltier [46]. Another promising alternative is the circulating fluidized bed combustor (see Fig. 4). It operates at lower temperatures (e.g., 1550–1700°F). However, it achieves high heat transfer efficiency by using conductive heat transfer mechanisms along with convective heat transfer mechanisms.

Heat exchanger design is the second area of improvement, where high temperature stainless steel (e.g., Iconel) systems have been proposed for regimes in the 1550–1650°F region. Ceramic heat exchangers also have been proposed and their commercial deployment appears most promising. Such heat exchangers can handle temperature regimes in the 2100–2300°F range.

Other than the heat exchanger, all of the elements in this cycle have been commercially proven. Consequently it is considered to be very close to commercialization.

2. Technoeconomic Analysis of the Air Turbine Cycle. The economic analysis of the downheat Brayton cycle depends upon establishing a representative system, calculating a heat balance for that system, and then establishing order of magnitude costs for such a unit. This economic analysis also depends upon recognizing that the costs attributable to

cogeneration are incremental costs above and beyond those
expenditures associated with steam raising. Cogenerated
power is a byproduct of process steam production in the FPI
[60].

The heat balance for one air turbine based cogeneration
cycle is shown in Fig. 15. That heat balance illustrates that
the air turbine actually produces less than 50% of all power
generated when electricity production is maximized.

Order of magnitude cost estimates have been developed for
the system presented in Fig. 15. Conceptual capital costs
are presented in Table IX while operating and maintenance
costs are presented in Table X. Further, both tables present
cost allocations recognizing the dominance of steam in the
energy requirements of the chemical pulp mill.

The technoeconomic analysis of this system is presented
in Table XI. The downheat Brayton cycle, which is a reversible innovation, is imminently ready for commercial deployment.
It can compete in any region where electricity can be sold at
a levelized price of about 55 mills/kWh. It is apparent that
systems with higher power/process heat ratios will be developed in the FPI.

B. Cattle Food Supplements

The production of cattle foods from wood is a far more
risky proposition than the introduction of the Brayton cycle
into wood fired power generation. Producing cattle food on a
large scale involves opening up new markets as well as developing new technologies. At the same time, there is a protein shortage in the world. Further, while the U.S. feeds
soybeans and grains to cattle in order to provide a high protein meat diet, most of the world eats such sources of
protein directly.

Animal food oriented crops have high economic values, as
is shown in Table XII. Of these soybeans are most interesting since they are used primarily as animal food in the U.S.
Soybeans contain 40% crude protein and are therefore useful
as cattle food. Currently some 68 million acres of U.S.
cropland are used to produce about 50×10^6 tons of soybeans
annually, of which about 30×10^6 tons are used in the U.S.
Some 20×10^6 tons are exported, principally to countries
where soybeans are human food [66].

Given the protein content of soybeans, the levelized cost
per pound of protein is 85¢/lb assuming a 30-year project
life, a 9.65% annual rate of escalation, and a 15% discount
rate. This 85¢/lb rate establishes the economic competition
for wood based cattle food.

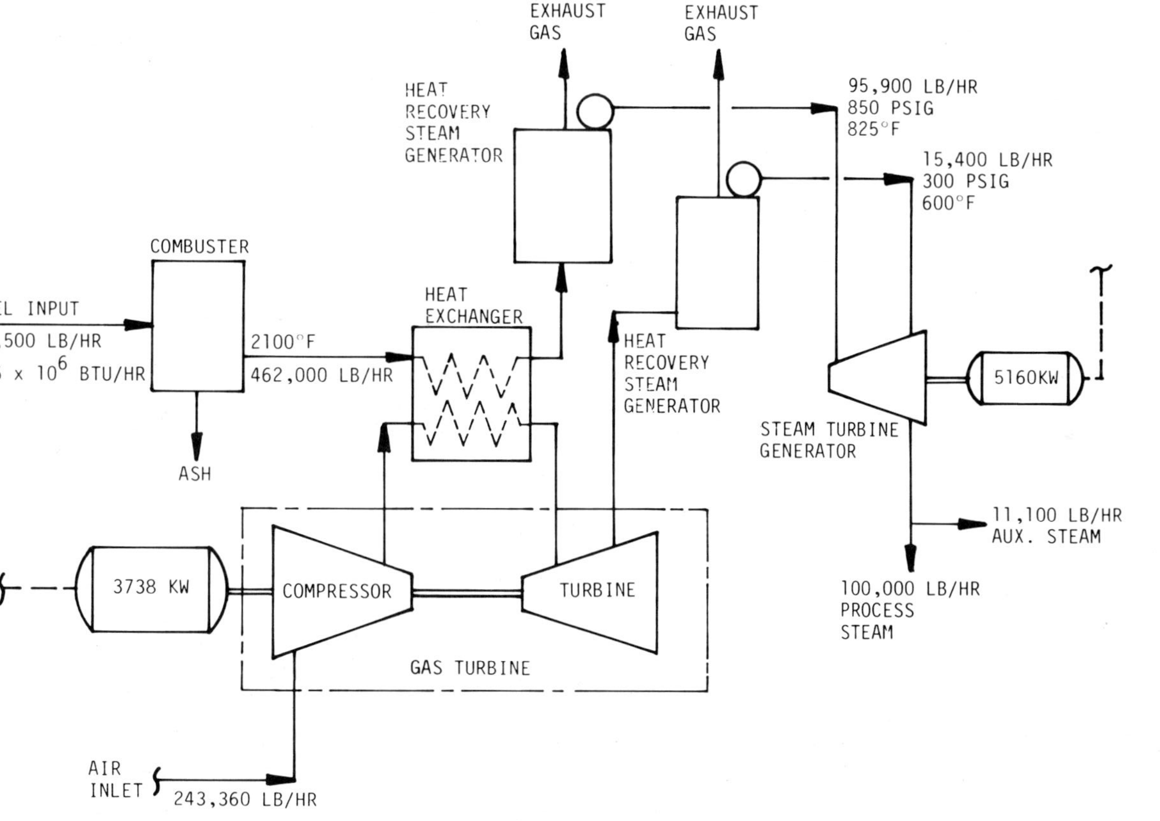

FIGURE 15. Heat balance of an air turbine or downheat Brayton cycle system. Source: Tillman, Schnorr, and Sale [63].

TABLE IX.   *Conceptual Estimate of the Capital Cost Associated with 8900 kW Downheat Brayton Cycle System (1984 $ \times 10^3$)*

Item	Equipment	Installation	Total
Material handling	930		
Gasifier/combustors[a]	2,240		
Heat exchanger	600		
Air turbine	1,520		
Waste heat boilers (2)	610		
Turbine generator[b]	1,400		
Other mechanical[c]	1,300		
Other electrical	800		
Other civil[d]	1,020		
Total directs	10,420	5,610	16,030
Engineering @ 10%			1,600
Construction management @ 10%			1,600
Working capital @ 5%			800
Contingency @ 5%			800
Total			20,830
Chargeable to power (Net of steam)[e]			10,100

[a] Lamb-Cargate cells.

[b] Assumes 4 MW turbine.

[c] Includes baghouse.

[d] Assumes no building required for combustor.

[e] Steam system approximated as the sum of items 1, 2/3 of 2, 5, 1/2 of 7, 8, and 9. This sum is divided by 0.65 and then multiplied by 1.3 to equal 10.7×10^6.

TABLE X.   Annual Operating and Maintenance Costs for 8900 kW Downheat Brayton Cycle System (1984 $ $\times$ 10^3)

	Total	Chargeable to steam	Chargeable to power
Fixed costs			
Labor[a]	*540*	*390*	*150*
Supplies and other	*75*	*50*	*25*
Maintenance @2.5% of capital	*521*	*268*	*253*
Variable costs (8000h/y)			
Fuel @ $2/Btu $\times$ 10^6	*3,936*	*3,232*	*704*
Total	*5,072*	*3,940*	*1,132*

[a]*$30,000/worker.*

1.   Producing Cattle Foods From Wood.   While experiments in the production of cattle food from wood date back to the 1920s, experimentation in chemical conversion of wood to cattle food became more prominent during the 1970s. At that time it was found that aspen is particularly susceptible to treatment by such reactants as gaseous or liquid ammonia, sodium hydroxide, or sulfur dioxide. Typical treatments and responses in cattle food production are shown in Table XIII.

The values for liquid ammonia are particularly interesting. Aspen chips treated with liquid ammonia contain 9% crude protein. The ammonia not only makes the carbohydrate more accessible, it increases the protein content of the feed.

2.   Economic Analysis of Cattle Food Manufacture.   A system for manufacturing cattle food using the liquid ammonia treatment, where yields are 100% on a mass basis, can be postulated. Costs can be developed based upon similarity between this equipment and a batch digester based kraft mill. Capital costs for such a system are shown in Table XIV. Operating and maintenance costs are shown in Table XV.

The data in Tables XIV and XV have been converted into a technoeconomic analysis using a pioneer plant (nominal) discount rate of 25%, and a mature technology discount rate of 15%. This analysis is presented in Table XVI. What is important about Table XVI is that the cost of wood is the

TABLE XI. Technoeconomic Analysis of Downheat Brayton Cycle System (Basis: Costs Chargeable to Power)

Parameter	Value
Capital investment	10.1×10^6
Accelerated cost recovery schedule	5 yrs
Annual O & M cost	1.132×10^6
Annual output	71.2×10^6 kWh
Project life	30 yrs
Discount rates (Nominal)	
Risk Adjusted	19%
Mature	15%
Inflation expectation	5%
RESULT	
Net present value (electricity) valued @ 5¢/kWh)	
Risk adjusted discount rate	3.50×10^6 (3.74×10^6 Real)
Mature discount rate	7.02×10^6 (6.48×10^6 Real)
Levelized production price	
Risk adjusted discount rate	5.4¢/kWh
Mature discount rate	4.9¢/kWh
[Purchased electricity][a]	7.5¢/kWh

[a] *Base values: 1984 price = 5¢/kWh; discount rate = 15%; inflation = 5%.*

dominant cost. At $50/O.D. ton, which is essentially a pulp chip value, the production of cattle food is not particularly attractive today. If that wood cost were reduced to fuel value, however, it would be priced at $35/O.D. ton. On that basis the production of cattle food from aspen could be competitive with soybeans.

It should be cautioned that this analysis is extremely simplistic, based upon some very preliminary information.

TABLE XII. Economic Values for Certain Commercial Crops

Crop	Current price $/BU	$/Ton	Levelized price[a] $/BU	$/Ton
Soybean	8	290	19	680
Corn	4	140	9	330
Wheat	5	180	12	420
Sorghum	4	140	9	330

[a]*9.7% rate of inflation as shown in the Statistical Abstract [66], 15% discount rate, and 30-year life.*

TABLE XIII. Typical Treatments and Responses in Producing Cattle Food from Wood

	Treatment			
Chemical	NH_3	NH_3	NaOH	SO_2
Temperature (°F)	85	130	N/A	250
Time (min)	60	30	60	120
Product digestibility[a] (%)	51[b]	48[b]	48[b]	63[b]

[a]*Untreated wood ranges from 0-33%, with aspen having 9-33%; alfalfa is 25%; hay is 50%.*

[b]*Aspen substrate.*

Source: Baker et al. [4].

TABLE XIV. Conceptual Capital Cost Estimate of a 1000 Ton/Day Cattle Food Facility (1984 $ × 10³)

Item	Cost
Wood yard	20,600
Materials handling	6,400
Chip treatment and washing	25,900
Pollution control	17,200
General and other	15,100
Subtotal	85,200
Engineering @ 10%	8,500
Construction management @ 10%	8,500
Working capital @ 5%	4,300
Contingency @ 5%	4,300
Total	110,800

TABLE XV. Conceptual Annual Operating Costs for a 1000 Ton/Day Cattle Food Facility (1984 $ × 10³)

Cost category	Cost
Fixed operating and maintenance	
Labor	5,200
Maintenance	2,800
Supplies	100
Variable operating and maintenance	
Wood ($50/O.D. ton)	17,500
Chemicals ($20/ton)[a]	1,400
Total	27,000

[a] Assumes 14,000,000 lb NH_3 @ $.10/lb as makeup.

TABLE XVI. Technoeconomic Analysis of Cattle Food Production Facility (Basis: 1000 Ton/Day Facility)

Parameter	Value
Capital investment	110.8×10^6
Accelerated cost recovery schedule	5 yrs
Annual O & M cost	27×10^6
Annual output	350×10^3 tons
Project life	30 yrs
Discount rates (Nominal)	
Risk adjusted	25%
Mature	15%
Inflation rate	5%
RESULTS	
Levelized production rate	
Risk adjusted	$200/ton ($110/ton Real)
Mature technology	$170/ton ($ 80/ton Real)
Levelized production price/lb protein	
Risk adjusted	$1.10
Mature technology	$0.90

Despite such limitations, these data point to foodstuff production as a highly intriguing new market for the FPI. It is a market of substantial size. It is also a market where wood has a significant advantage over agricultural crops in its ability to be stored "on the stump." Cattle food and human food offer significant opportunities for FPI growth.

V. CONCLUSION

The production of energy, chemicals, and specialty products is worth some $6-7 billion/yr to the Forest Products Industry. This value is obtained either from the sale of products (e.g., fuel, dissolving pulp, DMSO) or from the

avoiding of purchases of fossil fuels and electricity.
Energy, chemicals, and related products also provide the most
immediate and long term growth opportunities for the FPI.

Three forces unify the areas of energy, chemicals, and
related products production: (1) their dependence on pulp
mill byproducts as sources of fuel (e.g., kraft black liquor)
or chemical feedstocks (e.g., tall oil); (2) their dependence
on specialized products, markets, and economic conditions;
and (3) their reliance upon chemical rather than physical
characteristics of the raw material.

The economic importance of wood energy, chemical, and
related products has led to numerous developments in production technologies. Such developments include the improvement
of pile burning and fluidized bed combustion systems, wood
gasification, and electricity generation.

In addition to the recently commercialized technologies
such as the circulating fluidized bed, numerous energy and
chemical production systems have been proposed for development and deployment. Proposed energy systems include alternative cogeneration cycles such as the downheat Brayton cycle,
the wood fired gas turbine, and internal combustion reciprocating engine systems. Chemical systems proposed include
synthesis gas based schemes designed to produce methanol,
acetic acid, ethylene, ethylene glycol, and the entire range
of C_1 chemicals. Systems proposed also include deoxygenation
and hydrogenation systems necessary for synthesis and production of aromatic chemicals. Novel processes also have been
proposed for making foodstuffs from wood. Such systems can
produce either cattle food or human food.

The production of energy, chemicals, and related products
affords the FPI a window at new and expanding markets. These
new and expanding markets, such as the production of protein
containing cattle food, stand in stark contrast to the markets for such traditional products as lumber and plywood.
There is no vast overcapacity impeding the growth of wood
based energy and chemicals. There is no market stagnation.
Manufacturing many of these products does not require over-
coming the severe cost/price squeeze associated with the production of kraft or sulfite pulps (although that statement
does not hold for dissolving pulps).

The production of energy, chemicals, and related products
from wood has grown as a result of the restructuring of
fossil energy prices that occurred from 1973 to 1979. Further it resulted from the growing shortage of food in the
world. These events created an economic climate that supports higher risk investments such as advanced combustion
technologies. It is expected that this climate will continue
and will support investments in novel cogeneration, chemical
synthesis, and feedstuff manufacturing processes.

The area of energy, chemicals, and related products may indicate one of the strongest capital investment orientations of FPI expansion. If such is the case, the Forest Products Industry of the year 2000 will be substantially different from the FPI of today. It will be more oriented toward engineered products, and products dependent upon the chemical characteristics of the feedstock. It will be more oriented towards new technologies capable of yielding positive net present values with risk adjusted discount rates. It will seek out and penetrate growth markets.

The Forest Products Industry faces enormous challenges in seeking to change from an old growth to a managed growth resource base, and in seeking to change from a resource orientation to a market orientation. It can meet those challenges by maximizing the economic efficiency of its primary processing units—its kraft pulp mills. It can meet that challenge by engineering the most efficient and cost competitive lumber and board products. It can meet that challenge by focusing on such growth markets as energy, chemicals, and foodstuffs. New and potentially attractive technologies exist in all of these areas.

REFERENCES

1. Anderson, J. E. 1975. The Oxygen Refuse Converter— A System for Producing Fuel Gas, Oil, Molten Metal and Slag From Refuse. Union Carbide Corp., New York.

2. Anon. 1984. Concord—A City Heated With Wood. *Biologue* *1*(2):1.

3. Aquilo, A., Alder, J. S., Freeman, D. N., and Voorhoeve, R. J. H. 1983. Focus on C_1 Chemistry. *Hydrocarbon Processing 62* (3):57-65.

4. Baker, A. J., Millett, M. A., and Satter, L. D. 1975. Wood and Wood-Based Residues in Animal Feeds. *In* "Cellulose Technology Research" (A. F. Turbak, ed.). American Chemical Society, New York.

5. Barker, R. G. 1978. Potential for Chemically Induced Lightwood. *In* "Complete Tree Utilization of Southern Pine," Proc. (C. W. McMillin, ed.). Forest Products Research Society, Madison, Wisconsin.

6. Bliss, C., and Blake, D. O. 1977. Silvicultural Biomass Farms: Conversion Processes and Costs. Mitre Corp., McLean, Virginia.

7. Boyd, J. *et al.* 1975. Problems and Legislative Opportunities in the Basic Materials Industries. National Materials Advisory Board, National Academy of Sciences, Washington, D.C.

8. Bratt, L. C. 1979. Wood-derived Chemicals: Trends in Production in the U.S. *Pulp and Paper 53*(6):102-108.

9. Dalvi, A. D., and Burnett, T. C. 1982. Development of Technology for the Fluid Bed Gasification of Waste Wood With Pure Oxygen. The Sixth International Industrial Wood Energy Forum, Proc. Forest Products Research Society, Madison, Wisconsin.

10. Donath, E. E., and Hoering, M. 1977. Early Coal Hydrogenation Catalysis. *Fuel Processing Technology 1*:3-20.

11. Easterly, J. L., and Saris, E. C. 1984. A Survey of the Use of Biomass as a Fuel to Produce Electricity in the United States. Energy Technology XI: Applications and Economics. Government Institutes, Inc., Washington, D.C.

12. Ebasco Services Incorporated. 1982. Hulett, Wyoming Wood Waste to Electricity Feasibility Study. For Wyoming State Forestry Division. ESI, Bellevue, Washington.

13. Ebasco Services Incorporated. 1984. Feasibility Study: Wood Gasification Based Cogeneration Facility at the Wind River Nursery. For U.S. Department of Energy (as Administered by the Bonneville Power Administration). ESI, Bellevue, Washington.

14. Ebasco Services Incorporated. 1984. Biomass Fuel Supply Study. For Seattle City Light. ESI, Bellevue, Washington.

15. Ebasco Services Incorporated. 1984. Conceptual Study for Wood Residue Fired Power Plants. For Seattle City Light. ESI, Bellevue, Washington.

16. Emert, G. H., Katzen, R., Frederickson, R. E., Kaupisch, K. F., and Yeats, C. E. 1983. Update on the 50 T/D Cellulose-to-Ethanol Plant. *Journal of Applied Polymer Science: Applied Polymer Symposium 37*:787-795.

17. Enzer, H., Dupree, W, and Miller, S. 1975. Energy Perspectives. U. S. Department of the Interior, Washington, D.C.

18. Feldman, H. F., Liu, K. T., Longanbach, J. R., Curran, L. M., and Chauhan, S. P. 1980. Conversion of Forest Residues to a Clean Gas for Fuel or Synthesis. TAPPI 63(5):83-87.

19. Foster-Pegg, R. W., and Garland, R. V. 1982. Atmospheric Fluid Bed Air Turbine Cogeneration System. Westinghouse Electric Corp., Concordville, Pennsylvania.

20. Frank, M. E. 1981. Future Changes in Methanol Use and Methanol Synthesis. Presented at the American Chemical Society Annual Meeting, Aug. 25., New York City.

21. Frederick, W. J. 1984. The Energy Costs of Increased Organics Recovery for Chemical Byproducts in Kraft Pulp Mills. *In* "Progress in Biomass Conversion," Vol. 5. Academic Press, New York.

22. Fryling, G. R., ed. 1966. Combustion Engineering. Combustion Engineering Co., New York.

23. Gaden, E. L., Mandels, M. H., Reese, E. T., and Spano, L. A., eds. 1976. "Enzymatic Conversion of Cellulosic Materials: Technology and Applications." Wiley Interscience, New York.

24. Glesinger, E. 1949. "The Coming Age of Wood." Simon and Schuster, New York.

25. Goetzel, A., and Tatum, S. 1983. Wood Energy Use in the Wood Products Industry: What the Data Show. *In* "Progress in Biomass Conversion," Vol. 4. Academic Press, New York.

26. Goheen, D. W. 1971. Low Molecular Weight Chemicals. *In* "Lignins: Occurrence, Formation, Structure, and Reactions" (K. V. Sarkanen and C. H. Ludwig, eds.). Wiley Interscience, New York.

27. Hagen, K. G. 1977. Wood Fueled Combined Cycle Gas Turbine Power Plant. Presented at the California Energy Commission Workshop, July 15.

28. Hagen, K. G., and Berg, C. A. 1976. Wood Residue Fired Gas Turbine Cycle. *In* "Energy and the Wood Products Industry," Proc. Forest Products Research Society, Madison, Wisconsin.

29. Haggin, J. 1981. C_1 Chemistry Development Intensifies. *Chemical and Engineering News* 59(8):39-47.

30. Hajny, G. J. 1981. Biological Utilization of Wood for Production of Chemicals and Foodstuffs. U.S.D.A. Forest Service, Forest Products Laboratory, Madison, Wisconsin.

31. Hamrick, J. T. 1984. Installation of a 3-MW Wood Fired Gas Turbine at Red Rolling Springs, Tennessee. Pacific Northwest Bioenergy Systems: Policies and Applications Symposium, Portland, Oregon, May 10-11.

32. Hart, Fred C. and Associates. 1983. State-of-the-Art
    Survey of Wood Gasification Technology. EPRI, Palo Alto,
    California.

33. Herrick, F. W., Casebier, R. L., Hamilton, J. K., and
    Sandberg, K. R. 1983. Microfibrillated Cellulose:
    Morphology and Accessibility. *Journal of Applied Polymer
    Science: Applied Polymer Symposium 37*:797-814.

34. Hettich, B. V. 1980. Viscose Rayon—Technology for the
    Future. Fifth International Dissolving Pulp Conference,
    Proc., Vienna, Austria, Oct. 8-10.

35. Hokanson, A. E., and Katzen, R. 1978. Chemicals from
    Wood Waste. *Chemical Engineering Progress*, Jan., 67-71.

36. Hurley, P. 1978. Comparison of Mill Energy Balance
    Effects of Conventional, Hydropyrolysis, and Dry Pyroly-
    sis Recovery Systems. Institute of Paper Chemistry,
    Appleton, Wisconsin.

37. Irgolic, K. J., Calvert, R. A., Chapman, D. L., Clark,
    W. E., Gill, W. D., Krammer, A. P., Stranges, A. M., and
    Tooley, T. H. 1979. The World War II Germain Synfuels
    Program. *In* "Composition of Transportation Synfuels:
    R & D Needs, Strategies and Actions," Proc. U.S. Depart-
    ment of Energy, Washington, D.C.

38. Kahn, H. 1976. "The Next 200 Years." William Morrow,
    New York.

39. Katzen, R. and Associates. 1976. Chemicals From Wood
    Waste. U.S.D.A. Foest Service, Madison, Wisconsin.

40. Koch, P. 1982. Pulling Southern Pine With Taproots Is
    Economic Way to Boost Wood Harvest Per Acre 20 Percent
    and Increase Production of Naval Stores. The Sixth
    International FPRS Industrial Wood Energy Forum, Proc.
    Forest Products Research Society, Madison, Wisconsin.

41. Leppa, K. 1982. A Comparison of European and U.S.
    Combustion Systems Using Biomass. *In* "Progress in
    Biomass Conversion," Vol. 3. Academic Press, New York.

42. Levelton, B. H. and Associates. 1978. An Evaluation of
    Wood Waste Energy Conversion Systems. Environment of
    Canada, Vancouver, British Columbia.

43. Lin, S. Y. 1983. Lignin Utilization: Potential and
    Challenge. *In* "Progress in Biomass Conversion," Vol. 4.
    Academic Press, New York.

44. Meridian Corporation. 1983. Biomass Energy Technology
    Program Summary. U.S. Department of Energy, Washington,
    D.C.

45. Metz, C. B. 1980. Dissolving Pulps—The Future Economic Situation. Fifth International Dissolving Pulp Conference, Proc. Vienna, Austria, Oct. 8-10.

46. Mills, R. G., and Peltier, R. V. 1980. Wood Burning Indirectly Heated Gas Turbine/Cogeneration System for Use in the Pulp and Paper Industry. American Society of Mechanical Engineers, New York.

47. Moskowitz, S., Mullen, J., and Vanderlinden, S. 1983. A Coal Fired Gas Turbine Using an Air Cooled Fluidized Bed Combustor. American Society of Mechanical Engineers, New York.

48. Oaks, E. J., and Engstrom, F. 1982. Fluidized Bed Combustion Provides for Multifuel, Economical Cogeneration Systems. *Power Engineering,* March, 56-59.

49. Overend, R. 1979. Wood Gasification: An Overview. *In* "Hardware for Energy Generation," Proc. Forest Products Research Society, Madison, Wisconsin.

50. Paisley, M. A., Feldman, H. F., Applebaum, H. R., and Kim, B. C. 1982. Gasification of Forest Residues in a High Throughput Gasifier. Presented at the International Congress on Technology and Technology Exchange. Pittsburgh, Pennsylvania, May 3-6.

51. Post. 1984. Pulp and Paper Directory. Miller Freeman, San Francisco.

52. Probstein, R. F., and Hicks, R. E. 1982. "Synthetic Fuels." McGraw Hill, New York.

53. Puls, J., Ayla, C., and Dietrichs, H. H. 1983. Chemicals and Ruminant Feed From Lignocelluloses by the Steaming-Extraction Process. *Journal of Applied Polymer Science: Applied Polymer Symposium 37*:685-696.

54. Pyropower Corp. 1983. Pyroflow Circulating Fluidized Bed Combustion System. Pyropower Corp., San Diego, California.

55. Reed, T., and Markson, M. 1983. A Predictive Model for Stratified Downdraft Gasification of Biomass. *In* "Progress in Biomass Conversion," Vol. 4. Academic Press, New York.

56. Rowell, R. M., and Hokanson, A. E. 1979. Methanol From Wood: A Critical Assessment. *In* "Progress in Biomass Conversion," Vol. 1. Academic Press, New York.

57. Slinn, R. J. 1981. Developments in the U.S. Paper
    Industry as They Impact Naval Stores. Presented at the
    8th International Naval Stores Conference, New Orleans,
    Oct. 6.

58. Stone, R. N. 1977. A Critique of Completed Research on
    Paraquat-Induced Lightwood. Presented at Forest Biology
    Wood Chemistry Conference, Madison, Wisconsin, June.

59. Tillman, D. A. 1978. "Wood As An Energy Resource."
    Academic Press, New York.

60. Tillman, D. A., and Jamison, R. L. 1982. Cogeneration
    With Wood Fuels. *Fuel Processing Technology* 5:169-181.

61. Tillman, D. A., Rossi, A. J., and Kitto, W. D. 1981.
    "Wood Combustion: Principles, Processes, and Economics."
    Academic Press, New York.

62. Tillman, D. A., Rossi, A. J., and Simmons, S. O. 1982.
    Wood: Its Present and Potential Uses. For the U.S.
    Congress, Office of Technology Assessment. Envirosphere
    Co., Bellevue, Washington.

63. Tillman, D. A., Schnorr, R. F., and Sale, J. W. 1982.
    Alternative Cogeneration Systems Using Biomass Fuels.
    American Power Conference, Proc. Chicago, Illinois, April.

64. Tillman, D. A. 1983. Evaluating the Potential Contribu-
    tion of the Forest Products Industry to U.S. Energy
    Supply. *Journal of Applied Polymer Science: Applied
    Polymer Symposium* 37:593-615.

65. Turbak, A. F., Snyder, F. W., and Sandberg, K. R. 1983.
    Microfilbrillated Cellulose, a New Cellulose Product:
    Properties, Uses, and Commercial Potential. *Journal of
    Applied Polymer Science: Applied Polymer Symposium* 37:
    815-828.

66. U.S. Department of Commerce. 1984. Statistical Abstract.
    U.S. Government Printing Office, Washington, D.C.

67. U.S. Energy Information Administration. 1984. Annual
    Energy Outlook, 1983, With Projections to 1995. U.S.
    Department of Energy, Washington, D.C.

68. Van Der Heijden, S., Szladow, A. J., Barabas, M., and
    Sirianni, G. 1981. Wood Gasification System for Elec-
    tricity Production. American Society of Mechanical
    Engineers, New York.

69. Villesvik, G., and Tillman, D. 1983. Cofiring Dissimi-
    lar Solid Fuels: Some Fundamental and Design Considera-
    tions. American Power Conference, Proc. Chicago,
    Illinois.

70. Weisgerber, G. A., and Varma, A. 1979. Fluidized Bed Wood Gasifier. American Chemical Society, Washington, D.C.

71. Woodward, H. F. 1967. Methanol. *In* Kirk-Othmer "Encyclopedia of Chemical Technology," 2nd Ed., Vol. 13. John Wiley, New York.

72. Zylkowski, J. R., and Krippene, B. C. 1982. "Convert Conventional Boiler to Wood-Fired, Fluidized-Bed Unit." Energy Systems Guidebook. McGraw-Hill, New York, 23-27.

Appendix

GLOSSARY

*(Glossary of frequently used abbreviations, with
definitions given as appropriate.)*

Abbreviation	Definition/Explanation
A	Assets.
ACRS	Accelerated Cost Recovery System. This is the depreciation method defined in the 1981 tax act.
AFUDC	Allowance For Funds Used During Construction. Essentially this is the interest on the short term construction financing.
ASTM	American Society for Testing and Materials.
Beta (β_j)	This is the measure of systematic risk for the equity instruments (common stocks) of any publicly traded company.
Beta (β_u)	This is the measure of systematic risk for the equity instruments (common stocks) of any publicly traded company assuming no outstanding long term debt.
BF	Board Foot. 1 ft $\times$ 1 ft $\times$ 1 in. (nominal). Also see MBF, or 1,000 board feet.
C_1 chemicals	Chemicals made from synthesis gas (a mixture of carbon monoxide and hydrogen) such as methanol, ethylene, ethylene glycol, and formaldehyde are classified as C_1 chemicals.
CMP	Chemimechanical Pulping. This is the same system as CTMP pulping, applied to refiner mechanical pulping systems.
CRF	Capital Recovery Factor.
CTMP	Chemithermomechanical Pulping. Typically the addition of chemicals such as NaOH in the steaming operation, prior to the first stage refiner in the TMP mill is considered CTMP.

Abbreviation	Definition/Explanation
D	Depreciation. This is a non-cash operating expense used for capital recovery.
DCF	Discounted Cash Flow.
Def	Deflator. This is a factor used to remove expected future inflation expectations from cash flow estimates.
DGM	Dividend-Growth Model. This is an alternative to the CAPM for estimating the cost of equity capital.
DMSO	Dimethyl Sulfoxide. DMSO is an industrial solvent made from spent pulping liquor that has medicinal properties.
DR	Discount Rate.
FPI	Forest Products Industry.
hph	Horsepower hours. These are a measure of mechanical energy consumption.
I	Investment. The dollars spent on any capital project comprise the investment.
IRR	Internal Rate of Return. IRR is the rate of return earned by an investment as calculated by DCF analysis.
ITC	Investment Tax Credit.
k_d	The cost of funds acquired by debt instruments (e.g., bonds).
k_e	The cost of capital funds acquired by equity instruments (e.g., common stocks).
k_m	The average cost of equity capital.
k_{rf}	The cost of risk-free funds, typically treated as the cost of 90 day treasury bills.
kWh	Kilowatt hours.
L	Liabilities.
LCF	Loss Carry Forward. LCF is an accounting method for recovering previous tax losses through noncash charges on the income statement.
LPP	Levelized Production Price. More commonly this is referred to as Levelized Cost.

Abbreviation	Definition/Explanation
LVL	Laminated Veneer Lumber (also see PLV).
M	The premium for early availability and use of funds. This is a theoretical component of the discount rate.
MSP	Minimum Sale Price.
NATCF	Net After Tax Cash Flow. Those cash flows generated by an investment from both profits and non-cash charges (e.g., depreciation) are after tax cash flows.
NFPA	National Forest Products Association.
NFV	Net Future Value.
NPV	Net Present Value. NPV is the estimated present day increase in the wealth of any corporation or investor created by making any given investment. NPV, like NFV and IRR, is calculated using DCF techniques.
OE	Owners Equity.
O & M	Annual operating and maintenance costs for any given plant.
PGW	Pressurized Groundwood Pulping. PGW is one of the new mechanical pulping processes. It uses roundwood rather than chips as furnish.
PML	Project Market Line. The PML is a representation of the cost of capital as a function of the risk associated with a given project.
PLV	Parallel Laminated Veneer. Products made from veneer where the veneer grain is always laid up in a parallel direction are PLV products (also see LVL).
PURPA	Public Utilities Regulatory Policies Act. This is the Legislative part of the National Energy Plan stating that utilities must purchase cogenerated power at their avoided cost. This law has been tested and found to be constitutional.
R	Risk.
RADR	Risk Adjusted Discount Rate.
RMP	Refiner Mechanical Pulping. This is one of the mechanical pulping processes that uses chips rather than roundwood as furnish.

Abbreviation	Definition/Explanation
ROE	Return on Equity.
SGW	Stone Groundwood Pulping. This was the first wood pulping process invented, and it remains as one of the major mechanical pulping processes.
SML	Security Market Line. This is the representation of the cost of capital as a function of systematic risk for the corporation (β_j) derived from application of the CAPM.
TMP	Thermomechanical Pulping. This is the current state of the art in mechanical pulping, using residual chips as furnish, and producing the highest strength mechanical pulps.
TR	Tax Rate. This refers to the marginal corporate income tax rate.
WACC	Weighted Average Cost of Capital. This is typically considered to be the discount rate or hurdle rate for any company.
YTM	Yield to Maturity. This is the true yield on bond financing, and it represents the cost of debt capital.

Index